Grassland Biomes

GREENWOOD GUIDES TO
BIOMES OF THE WORLD

Introduction to Biomes
Susan L. Woodward

Tropical Forest Biomes
Barbara A. Holzman

Temperate Forest Biomes
Bernd H. Kuennecke

Grassland Biomes
Susan L. Woodward

Desert Biomes
Joyce A. Quinn

Arctic and Alpine Biomes
Joyce A. Quinn

Freshwater Aquatic Biomes
Richard A. Roth

Marine Biomes
Susan L. Woodward

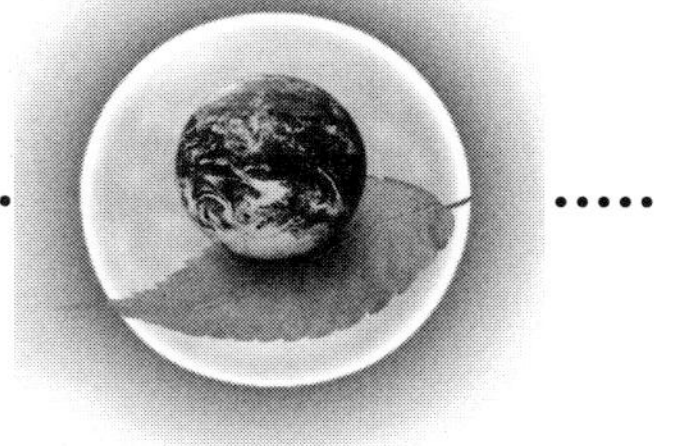

Grassland BIOMES

Susan L. Woodward

Greenwood Guides to Biomes of the World
Susan L. Woodward, General Editor

GREENWOOD PRESS
Westport, Connecticut • London

Library of Congress Cataloging-in-Publication Data

Woodward, Susan L., 1944 Jan. 20–
Grassland biomes / Susan L. Woodward.
p. cm. — (Greenwood guides to biomes of the world)
Includes bibliographical references and index.
ISBN 978-0-313-33840-3 (set : alk. paper) — ISBN 978-0-313-33999-8 (vol. : alk. paper)
1. Grassland ecology. 2. Grasslands. I. Title.
QH541.5.P7W66 2008
577.4—dc22 2008027508

British Library Cataloguing in Publication Data is available.

Library of Congress Catalog Card Number: 2008027508
ISBN: 978-0-313-33999-8 (vol.)
978-0-313-33840-3 (set)

First published in 2008

Greenwood Press, 88 Post Road West, Westport, CT 06881
An imprint of Greenwood Publishing Group, Inc.
www.greenwood.com

Printed in the United States of America

The paper used in this book complies with the Permanent Paper Standard issued by the National Information Standards Organization (Z39.48–1984).

10 9 8 7 6 5 4 3 2 1

Contents

Preface

This book describes and compares the major grassland regions of the world. Divided broadly into tropical and temperate biomes, each part of Earth dominated by grasses and other herbaceous plants is presented in terms of its climatic peculiarities, most prevalent soil types, characteristic structure of the vegetation, and typical plants and animals. A general global overview of each biome is followed by more detailed descriptions of its primary regional expressions on the different continents. Maps, diagrams, photographs, and line drawings enhance the readers' appreciation of the forms nature takes in various parts of the world and their understanding of why both similarities and differences may appear in separate parts of the same biome. Advanced middle-school and high-school students are the intended audience, but undergraduate students and anyone else interested in our planet's terrestrial environments will find material they can use.

A geographer's perspective will be evident: place matters. Location helps determine the assemblage of plants and animals found anywhere, in part, because basic environmental elements (such as temperature and precipitation patterns) are the products of the interplay between latitude (position north or south of the equator) and the circulation of the atmosphere. Position on a continent—near the coast or in the interior, on the windward or leeward side of a mountain range, for example—also matters as it influences rainfall amounts, winds, and temperatures. Location, however, has not been static nor has Earth's surface been unchanging through geologic time. The dynamics of Earth history has greatly influenced the distribution patterns of the life we see today. Each species had a place of origin, as well as a time of origin, and most later dispersed out of that place as land connections or barriers, themselves ever changing, permitted. Each species gained its own unique distribution area. Some species, or genera, or families became widespread or cosmopolitan, showing up in grasslands everywhere. Some stayed close to their original home and became what we now refer to as endemic species found only in one place or region.

As Earth changed and as plants and animals moved, populations adapted to the strongest prevailing environmental conditions of the day, were they climatic, edaphic, biological, disturbance factors, or some combination of all. If they couldn't adapt, they either never became established in new locations or went extinct. So time and space—geological and evolutionary history and physical geography—went hand in hand to determine which forms of life assembled together in any given region. Once together, they had to adapt to each other as well as to the physical environment and its limitations or opportunities. In widely separated locations but under similar environmental pressures, unrelated lifeforms adopted similar strategies for survival and evolved similar shapes, sizes, and behaviors. Thus, it is possible make generalizations about the plant and animal life in a given area without getting lost in the details of which individual species are actually present. This type of generalization is at the heart of the biome concept. One studies the vegetation and looks at the form or structure of the overall community. Growthforms, layering of foliage, and the spacing of plants indicate adaptations to the environment and give a distinctive appearance to the natural landscapes of each biome. The emphasis in this book is on entire communities and their arrangement horizontally on the surface as well as vertically from the ground to the top of the plant cover. Vegetation profiles or cross-sections are used in addition to photographs to help the reader visualize the patterns life makes on different parts of the Earth's surface.

Species composition is not neglected, because it forms an integral part in the distinctiveness of each regional expression of a biome as well as between biomes. What species are actually found in which locations is also important to know if we are to conserve a major part of the great diversity of lifeforms, habitats, and ecosystems that remain on the Earth but that are quickly being degraded or eliminated. Complete species lists are not given, but common and unique plants and animals are identified, especially if they give character to a particular region.

The notion that place matters directs the organization of this book and makes it different from many other reference books that deal with biomes. Rather than give an encyclopedic treatment of the organisms living in the world's grasslands in alphabetical order, a regional exploration of life's variety unfolds.

Reading a book about plants and animals is no substitute for actually seeing them in their natural habitats and observing how they fit into the total landscape or how they behave as individuals or members of social groups. I have most recently traveled to the highveld of South Africa, the dry savannas of the Kalahari, and the moist savannas of Kruger National Park. The freshness of the experience and the reawakening of wonder last a long time. I can only hope that the descriptions and illustrations in this volume peak the readers' interest enough to make them want to go see it all for themselves and better understand the values of conserving our natural heritage.

I would like to thank Kevin Downing of Greenwood Press for his insights and constant support in bringing this project to fruition. Not only are Jeff Dixon's illustrations a major contribution to the goals of the book, but he was a wonderfully cooperative collaborator in its production. Bernd Kuennecke of Radford University's Geography Department prepared the excellent maps that guide the reader to the regions described for the Temperate Grassland and the Tropical Grassland biomes. Joyce Quinn explored southern Africa's savannas and deserts with me and always had insightful questions and suggestions that enhanced not only this volume but also other Greenwood Guides to the Biomes of the World as well. My deepest appreciation goes to all these people.

Finally, I would like to dedicate this book to the memory of my father, Appleton C. Woodward, who first took me bird-watching in the woods of New England as a small child and kindled my lifelong interest in nature and geography.

Blacksburg, Virginia
January 2008

How to Use This Book

The book is arranged with a general introduction to grassland biomes and a chapter each on the Temperate Grassland Biome and the Tropical Savanna Biome. The biome chapters begin with a general overview at a global scale and continue to regional descriptions organized by the continents on which they appear. Each chapter and each regional description can more or less stand on its own, but the reader will find it instructive to investigate the introductory chapter and the introductory sections in the later chapters. More in-depth coverage of topics perhaps not so thoroughly developed in the regional discussions usually appears in the introductions.

The use of Latin or scientific names for species has been kept to a minimum in the text. However, the scientific name of each plant or animal for which a common name is given in a chapter appears in an appendix to that chapter. A glossary at the end of the book gives definitions of selected terms used throughout the volume. The bibliography lists the works consulted by the author and is arranged by biome and the regional expressions of that biome.

All biomes overlap to some degree with others, so you may wish to refer to other books among Greenwood Guides to the Biomes of the World. The volume entitled *Introduction to Biomes* presents simplified descriptions of all the major biomes. It also discusses the major concepts that inform scientists in their study and understanding of biomes and describes and explains, at a global scale, the environmental factors and processes that serve to differentiate the world's biomes.

The Use of Scientific Names

Good reasons exist for knowing the scientific or Latin names of organisms, even if at first they seem strange and cumbersome. Scientific names are agreed on by international committees and, with few exceptions, are used throughout the world. So everyone knows exactly which species or group of species everyone else is talking about. This is not true for common names, which vary from place to place and language to language. Another problem with common names is that in many instances European colonists saw resemblances between new species they encountered in the Americas or elsewhere and those familiar to them at home. So they gave the foreign plant or animal the same name as the Old World species. The common American Robin is a "robin" because it has a red breast like the English or European Robin and not because the two are closely related. In fact, if one checks the scientific names, one finds that the American Robin is *Turdus migratorius* and the English Robin is *Erithacus rubecula.* And they have not merely been put into different genera (*Turdus* versus *Erithacus*) by taxonomists, but into different families. The American Robin is a thrush (family Turdidae) and the English Robin is an Old World flycatcher (family Muscicapidae). Sometimes that matters. Comparing the two birds is really comparing apples to oranges. They are different creatures, a fact masked by their common names.

Scientific names can be secret treasures when it comes to unraveling the puzzles of species distributions. The more different two species are in their taxonomic relationships, the farther apart in time they are from a common ancestor. So two species placed in the same genus are somewhat like two brothers having the same father—they are closely related and of the same generation. Two genera in the same family might be thought of as two cousins—they have the same grandfather, but different fathers. Their common ancestral roots are separated farther by time. The important thing in the study of biomes is that distance measured by time often means distance measured by separation in space as well. It is widely held that new species come about when a population becomes isolated in one way or another from the rest of its kind and adapts to a different environment. The scientific classification into genera, families, orders, and so forth reflects how long ago a population went its separate way in an evolutionary sense and usually points to some past environmental changes that created barriers to the exchange of genes among all members of a species. It hints at the movements of species and both ancient and recent connections or barriers. So if you find two species in the same genus or two genera in the same family that occur on different continents today, this tells you that their "fathers" or "grandfathers" not so long ago lived in close contact, either because the continents were connected by suitable habitat or because some members of the ancestral group were able to overcome a barrier and settle in a new location. The greater the degree of taxonomic separation (for example, different families existing in different geographic areas), the longer the time back to a common ancestor and the longer ago the physical separation of the species. Evolutionary history and Earth history are hidden in a name. Thus, taxonomic classification can be important.

Most readers, of course, won't want or need to consider the deep past. So, as much as possible, Latin names for species do not appear in the text. Only when a common English language name is not available, as often is true for plants and animals from other parts of the world, is the scientific name provided. The names of families and, sometimes, orders appear because they are such strong indicators of long isolation and separate evolution. Scientific names do appear in chapter appendixes. Anyone looking for more information on a particular type of organism is cautioned to use the Latin name in your literature or Internet search to ensure that you are dealing with the correct plant or animal. Anyone comparing the plants and animals of two different biomes or of two different regional expressions of the same biome should likewise consult the list of scientific names to be sure a "robin" in one place is the same as a "robin" in another.

1

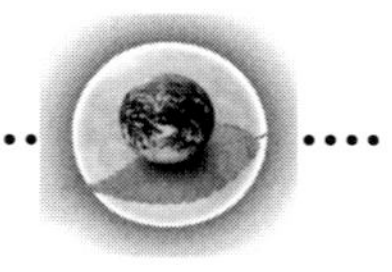

Introduction to Grassland Biomes

Two world biomes have grasses as the main type of plant; one is in the tropics and the other is in the middle latitudes (see Figure 1.1). The dominance of grasses, however, is just about the only thing the two biomes have in common. They have different climates (though limited precipitation is a factor in both), different soils, different animals, different nongraminoid growthforms, and even different types of grasses (see Table 1.1).

This volume gives information on both the Temperate Grassland Biome and the Tropical Savanna Biome. Each biome is described, beginning with a general global overview in which the following factors are considered:

- the geographic location of the biome,
- the general climatic conditions and other controls under which this vegetation naturally prevails,
- the common growthforms and structure of the vegetation of the biome,
- the major soil-forming processes and types of soils that develop as a result of the interaction between the climate and the vegetation,
- the common adaptations and kinds of animals living in the biome, and
- an indication of the current condition of and threats to the world's grasslands.

After the general global overview has been presented, each major part of the biome is described separately, arranged according to the continent on which it occurs. Details on actual location, climate conditions, soil types, and plant and animal species in each geographic region are presented in these later sections.

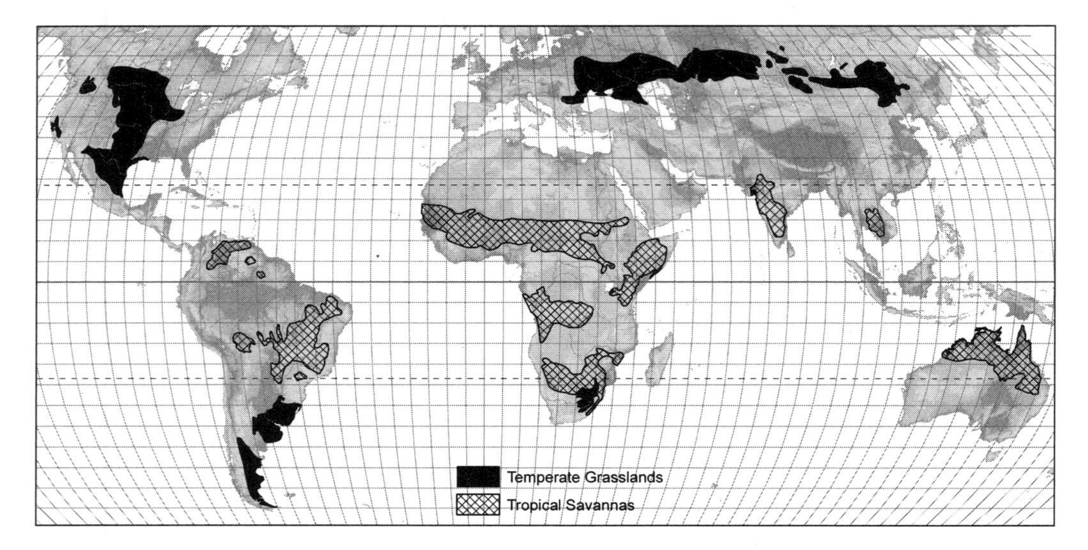

Figure 1.1 World distribution of the two major grassland biomes, the Temperate Grassland Biome and the Tropical Savanna Biome. *(Map by Bernd Kuennecke.)*

Grass refers a taxonomic unit, a family of plants now usually called Poaceae. Grasses have distinct physical characteristics shared with rushes, sedges, and bulrushes that are collectively referred to as the graminoid growthform. The term graminoid is derived from one of the scientific names applied to the grass family, Graminae. Other graminoids belong to different families: Rushes make up the family Juncaceae, whereas sedges and bulrushes are in the family Cyperaceae.

About Grass

True grasses are flowering plants. Their small flowers, called florets, have no petals and are often overlooked. Thousands of different species make up this very large and widespread family. Some are perennials living two or more years; others are annuals and complete their life cycles in one year or less.

Common features of any grass are shown in Figure 1.2. The mature grass plant consists of upright stems or culms. The culm has a series of solid joints or nodes covered by the basal sheaths of the leaves. The upper parts of the leaves, the blades, are arranged so they project alternately from opposite sides of the culm. Between nodes, in the internodes, the stem is either hollow or filled with a spongy material called pith. Rushes, sedges, and bulrushes have no nodes. The stems of rushes are round, while those of sedges and bulrushes are triangular. Many grasses produce stolons, aboveground stems that lie along the ground. New roots and daughter plants called tillers form at the nodes of stolons. Underground stems called

Table 1.1 Comparison of Temperate Grassland and Tropical Savanna Biomes

	Temperate Grasslands	Tropical Savannas
Location	Middle latitudes; interior of large continents or rainshadow of major mountain ranges; between forest and desert biomes	Tropics; north and south of Tropical Rainforest Biome
Temperature controls	Mid-latitude seasonality aggravated by continentality; intermediate elevations	Tropical latitude; low elevations
Temperature pattern (annual)	Wide range of temperatures; below freezing temperatures common in winter	Little month-to-month variation: no frost
Precipitation controls	Summer convectional storms; shifts in Polar Front; Asian monsoon; rainshadow effect	Seasonal shift of ITCZ
Precipitation totals	10–20 in (250–750 mm)	20–60 in (500–1,500 mm)
Seasonality	Based on hot summers and cold winters	Based on high-sun rainy season and low-sun dry season
Climate type	Semiarid	Tropical wet and dry
Dominant growthforms	Graminoids and forbs	Graminoids and trees
Photosynthetic pathway	C_4 and C_3	C_4
Dominant soil-forming process	Calcification	Laterization
Major soil order	Mollisol	Oxisol
Soil characteristics	Nutrient-rich, high humus content; neutral to slightly alkaline; black or brown	Nutrient-poor, heavily leached, acidic; red
Typical mammals	A few large grazing ungulates; few carnivores; colonial burrowing rodents	Africa: Many large herd-forming ungulates, both grazers and browsers; many large cats and other carnivores Elsewhere: Burrowing small mammals, mostly rodents
Biodiversity	Low to moderate	High
Age	Recent: Post-Pleistocene origins	Ancient: Tertiary origins
Current status	Modified by grazing and destroyed through conversion to crop cultivation since 1800s; more recently urbanization is a threat. The few remaining areas are often degraded from overgrazing and fire protection. Restoration projects under way.	Altered by thousands of years of burning and livestock grazing in Africa and Asia. There and elsewhere threatened by overgrazing, rapidly expanding crop cultivation, desertification, and urbanization. Large preserves established, at least on paper.

Note: ITCZ = Intertropical Convergence Zone.

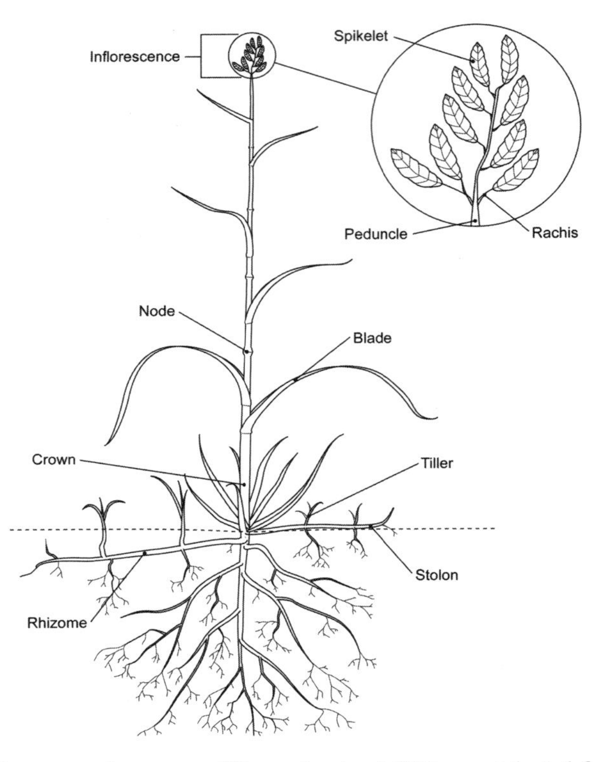

Figure 1.2 Main parts of a grass. *(Illustration by Jeff Dixon. Adapted from Oregon State University Forage Information Service 2000.)*

rhizomes also will root and send up new tillers. Both are means of vegetative reproduction that let the grass plant spread and form a turf or sod.

Grasses are often distinguished as either sod-forming species or bunchgrasses (see Figure 1.3). Bunchgrasses have their aboveground shoots clumped around a central stem and do not produce stolons or rhizomes. Bunchgrasses depend on seeds for reproduction and dispersal.

The base of a perennial grass, its crown, connects roots to stems. The crown survives the nongrowing season and produces renewal buds that will form new culms, tillers, roots, rhizomes, and stolons. It is the key part of the plant that allows for regrowth not only each year but also after disturbances such as grazing or fire. Some perennial grasses also store energy and nutrients in the crown area in organs called corms to boost regeneration in the coming growing season. Annual grasses do not have crowns.

Grasses are well adapted to a variety of environments that might overly stress trees. The upright stems and blades adapt grass to hot, high-sun conditions by keeping sunlight from striking the leaf at a high angle. This prevents destructive overheating of the photosynthetic surfaces. The crown allows perennial grasses to die

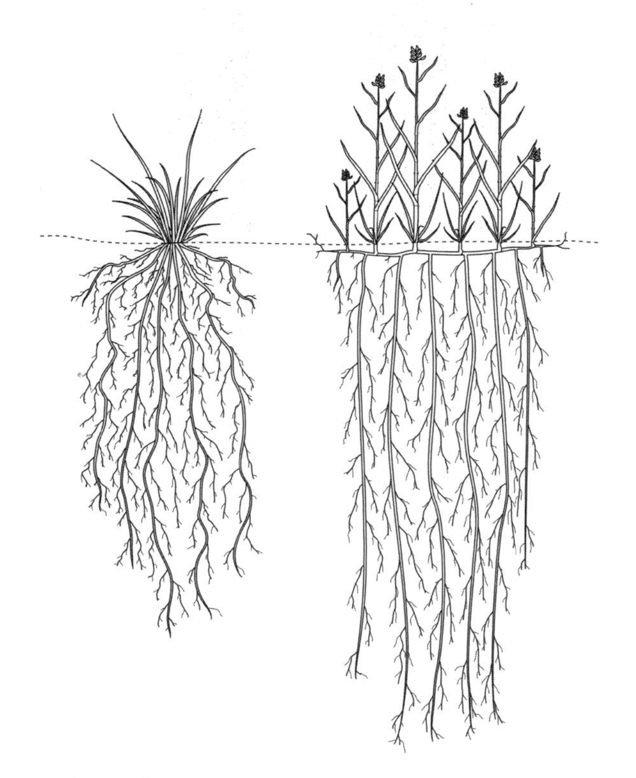

Figure 1.3 Comparison of the growthform of a bunchgrass (left) and a sod-forming grass (right). *(Illustration by Jeff Dixon. Adapted from Weaver 1968.)*

back during unfavorable conditions and then resprout when the temperatures or precipitation conditions permit growth. The position of the renewal bud at ground level, where it can become covered by dead blades or snow, offers protection against cold and often fire. The position of the renewal buds is also out of reach of most grazing animals (and lawnmowers), so cropping the grass does not kill it but actually stimulates growth.

The stiffness that holds grass blades upright comes from the cellulose of plant cell walls and from special bodies containing silica, the stuff of sand. These silica bodies, called phytoliths ("plant stones"), may be contained in special silica cells or accumulate between the walls of other cells. The phytoliths of each type of grass are distinctively shaped and are features that allow for identification of species under a microscope. This fact has allowed scientists to study the diets of grass-eating animals by examining plant remains in their feces.

Grasses had appeared by 55 million years ago during the Eocene Epoch. Animals adapted to the new indigestible and tooth-wearing forage by evolving new forms of digestion and long teeth able to withstand a lifetime of wear from eating these gritty plants. First the perissodactyls, the odd-toed ungulates or hoofed mammals (horses, rhinoceroses, and tapirs), came to dominance. Digestion was achieved by bacteria in a long caecum, a part of the digestive tract between the small intestines and the colon. This was not especially efficient, but animals could thrive by eating a large volume of coarse grasses and passing it quickly, though poorly digested, through their alimentary canals.

Later the even-toed ungulates, the artiodactyls, achieved greater digestive efficiency with the development of a multichambered stomach, including a rumen. Bacteria still played an essential role, but now their activity took place in a specialized fermentation chamber located in the foregut. The digestion process requires rechewing a bolus of food (the cud) and is slow, but less forage is required than amounts needed by nonruminants. Artiodactyls evolved into many species—especially in the cattle family (Bovidae)—and largely, though not completely, replaced the perissodactyls. The great success ungulates have had adapting to grass can still be seen on the world's grasslands today. Hoofed animals continue to be the main large mammals in most regions of both grassland biomes

Lagomorphs (rabbits and hares, pikes, and others) are a group of small mammals that eat grasses and are conspicuous members of the temperate grassland fauna of North America and Eurasia. They manage to extract energy and nutrients from grasses by passing the plant matter through their guts twice. In the first passage, freshly cropped grass is partially digested; undigested material is converted to moist fecal pellets and eliminated. The animal then eats these pellets and gets a second chance to remove the food value from their coarse forage. The final waste products are dry fecal pellets, which are not consumed.

Grasses are a geologically young and highly diversified family of plants whose members are adapted to a wide range of environmental conditions. Most are intolerant of shade, so closed-canopy woodlands and forests are generally devoid of grasses. Extreme aridity also limits grasses. Therefore, grasses tend to dominate in environments where, for one reason or another, trees are rare or widely scattered and where it is also not so dry that only desert shrubs can thrive. As a group, grasses are tolerant of a variety of temperature patterns, although the different species sort themselves out geographically according to temperature. Two modes of photosynthesis, known as the C_3 and C_4 pathways, respectively, have developed among the grasses, so that some grow best in cooler areas and others do better under hot conditions. Cool-season grasses or C_3 grasses differ in other ways from warm-season grasses or C_4 grasses as discussed in the sidebar on p. 7.

Grasses may be distinguished by their height. Tall grasses will be 6 ft (2 m) or more at maturity; mid-size grasses reach 4 ft (1.5 m) high, and short grasses stand less than 3 ft (1 m) tall.

Grassland Climates

The differences in the climates associated with tropical savannas compared with those of the temperate grasslands are key determinants of plant growth patterns. Climate includes both annual and seasonal precipitation amounts and seasonal temperature changes, if any. Tropical savannas occur in areas with a tropical wet and dry climate (Aw in the Koeppen climate classification), where 20–40 in (500–1,500 mm) of rain falls a year, but precipitation is concentrated in a roughly four-month period. In other words, a clear and lengthy dry season occurs during which many plants are unable to grow. Temperatures remain warm all year. Indeed, in the tropics, seasons are based on rain or no rain, and not on warm or cool temperatures. The rainy season occurs during high sun, when the Intertropical Convergence Zone (ITCZ) migrates poleward from the equatorial region. At this zone, the trade winds of both hemispheres meet and rise to produce rain. The high-sun period occurs during the summer of mid- and high-latitude places in a given hemisphere, so in the Northern Hemisphere the rainy season in most apt to be from May through August and in the Southern Hemisphere from November through February.

The temperate grasslands, on the other hand, are associated with mid-latitude semiarid climates (BS in the Koeppen climate classification system) in which precipitation is usually less than 20 in (500 mm) a year, but some may fall every month of the year. A peak in precipitation typically occurs in summer. Seasons in continental semiarid climate regions are distinguished by wide differences in temperature. Winters are apt to be bitterly cold and summers very hot (see Figure 1.4).

Cool-Season and Warm-Season Grasses or C_3 and C_4 Grasses

When plants fix carbon dioxide from the atmosphere during photosynthesis, they manufacture compounds based on carbon atoms. The initial product is an acid. It may have a 3-carbon structure or a 4-carbon structure depending on the chemical pathway used to change carbon dioxide to compounds plants can use. The number of carbon atoms involved in the pathway gives rise to the designation of C_3 or C_4 plants.

C_3 plants are also called cool-season plants because they grow best at moderate temperatures. Optimal temperatures are 65°–75° F (17°–23° C). Growth begins in early spring when soils temperatures reach 40°–45° F (4°–7° C) and may cease during the hottest parts of summer. Growth resumes during the shorter, cooler days of mid-latitude autumns.

The C_4 chemical pathway allows photosynthesis to take place at higher temperatures, so grasses that use this mode are referred to as warm-season grasses. C_4 grasses are most efficient at photosynthesis when temperatures are 90°–95° F (32°–34° C). They need soil temperatures of 60°–65° F (15°–18° C) to begin growth.

Grasses of the tropics and those growing in hot summer regions of the middle latitudes are typically C_4 grasses. C_3 grasses are abundant in the cooler parts of the temperate or middle latitude grasslands. Wheat and rye, both domesticated annual grasses, are C_3 species. Maize (corn) and sugar cane, also domesticated grasses, are C_4.

Grassland Soils

Climate plays a major role in soil formation so it is not surprising that major differences in the soils of these two biomes exist. Tropical savanna soils often have developed on ancient, stable surfaces that have been exposed to warm temperatures and at least moderate amounts of rain for perhaps millions of years. Most chemicals that could be dissolved and washed out of them (leached) have been, so they tend to have low fertility and concentrate oxides of iron and aluminum. The iron content gives these so-called oxisols a deep red color. The temperate grasslands formed after the last ice ages, only

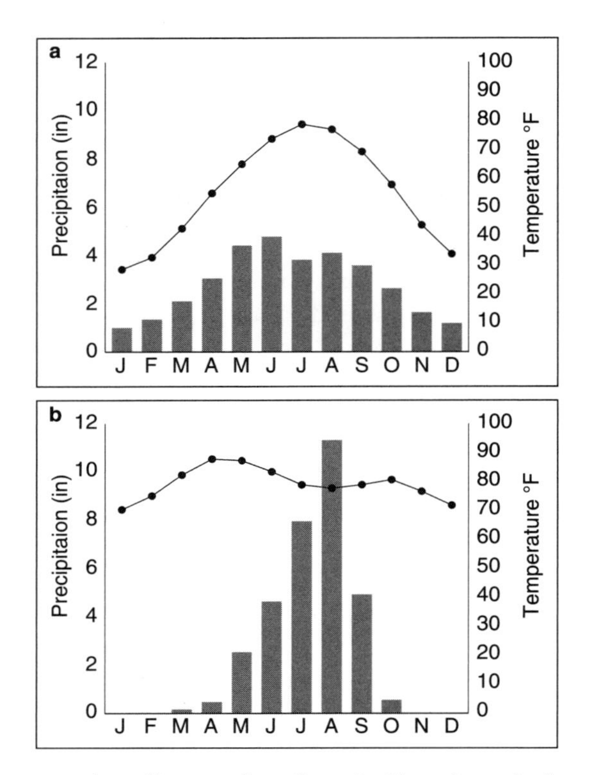

Figure 1.4 Representative climographs of semiarid and tropical savanna climates. (a) Topeka, Kansas, in the tall-grass prairie of North America; (b) Kano, Nigeria, in the Sudanian zone of the tropical savanna of Africa. *(Illustration by Jeff Dixon.)*

some 6,000–8,000 years ago. In the semiarid climate under which these soils developed, plant nutrients are not leached from the soil column but instead tend to concentrate in the middle layer (the B horizon) of soil into which grass roots penetrate. Cold winters allow for a seasonal slowing and cessation of the breakdown of dead and decaying plant material, so that nutrient-retaining humus collects in the topsoil (the A horizon). The result is a dark brown or even black, very fertile mollisol.

Human Impacts and Conservation Needs

The grass cover, the soils, and technology dictated the way the world's grasslands have been used by humans. Herding economies based on cattle, sheep, and goats have flourished in the Old World grasslands for millennia. Overgrazing changed the composition of the plant cover and created bare ground that allowed for the invasion of thorny shrubs and trees. Domestic livestock came to the New World grasslands after the first voyage of Christopher Columbus in 1492. Their spread onto both the temperate and tropical grasslands was delayed another few centuries

E-Grasses

Energy farming to produce biofuels is on the horizon to solve some environmental and economic problems. Nearly always the crop is a C_4 (warm-season) grass. The goal is to convert sugars or cellulose to ethanol in a sustainable agricultural system and to have a clean-burning replacement for petroleum and coal for use in transportation and power generation. Brazil has been using sugarcane, a tropical crop, this way for nearly two decades.

The primary e(nergy)-grass in the temperate United States is corn (maize), but it is not an efficient producer of ethanol. Only the kernels are used; during fermentation, their sugars change to alcohol. Growing corn requires almost as much gasoline as the ethanol that can be produced. So scientists and energy companies are experimenting with other potentially more efficient e-grasses, particularly giant miscanthus, wild cane, and switchgrass. In all three, the entire plant would be used because processing involves converting cellulose to biofuel, as is done with sugarcane.

Giant miscanthus (*Miscanthus* x *giganteus*), sometimes called elephant grass, is a tall manmade plant, a sterile hybrid of two Asian grasses. Wild cane or giant reed (*Arundo donax*) was introduced to the United States in the early 1800s from the Mediterranean region. Growing more than 20 ft (6 m) tall, this invasive grass would be grown for electricity generation, but at some ecological risk to the wetlands that it prefers. Switchgrass (*Panicum virgatum*), on the other hand, is a widespread, tall-grass prairie native. The hope is that it can be grown in a variety of areas marginal to food crop production, will provide a clean fuel with minor inputs of fossil fuels during its growth and harvest phase, and—if a market for it develops—keep the family farmer in business: E for excellent in this case!

until European settlement expanded into these less-than-hospitable environments in the 1800s. It was in 1837 that John Deere invented a steel plow that could cut through the thick sod of the prairies, steppes, and pampas of the temperate regions and made possible the cultivation of cereal grains, especially wheat and maize, on some of the world's richest soils. Today, large-scale cultivation of crops such as soybean is affecting tropical grasslands as well.

Human settlement has meant a change in the frequency and temperature of wildfires in both types of grassland. In the temperate grasslands, the grass fires that the natural vegetation depended on for regeneration were stopped. One consequence has been the invasion of woody plants. In the tropical grasslands, more often than not, the frequency of fires was increased by pastoralists to the detriment of some grasses. Again the composition of the plant community changed.

As a result of human activity, natural grasslands are rare today. Temperate grasslands generally survive only in small, managed patches. Tropical savannas are starting to face the same losses. Large preserves have been mapped out, but it remains to be seen how much of the biome can be conserved.

2

The Temperate Grassland Biome

General Overview

Natural vegetation dominated by perennial grasses and forbs is a feature of the mid-latitudinal regions of all continents (except Antarctica, which lies well outside the mid-latitudes) (see Figures 2.1 and 2.2). The Temperate Grassland Biome is associated with a semiarid climate and geographically occurs between temperate forests and deserts. The present assemblages of plants and animals date to after the Pleistocene, when modern climates were established, deposits of loess and glacial outwash formed, and the major wave of large mammal extinction that followed the last ice age nearly over.

Each continent has its own popular name for its part of the Temperate Grassland Biome. In North America, it is the prairie; in Eurasia, it is the steppe. South Americans usually refer to pampas and South Africans to the veld. The plants (and animals) of the North American and Eurasia sections of the biome are closely related, but they have been affected differently by Pleistocene and post-Pleistocene climate change, human occupation and use, wildfires, and grazing pressures from both wild and domesticated large mammals. The origins of South America's pampas and southern Africa's veld are still poorly understood. Fire may be implicated, because it is a necessary management tool used today to prevent the encroachment of woody plants.

Gradation in precipitation amounts occurs in most regional expressions of the biome and is mirrored in the zonation pattern of the vegetation and animal life in each. In some instances, such as in North America, longitudinal zones dominate

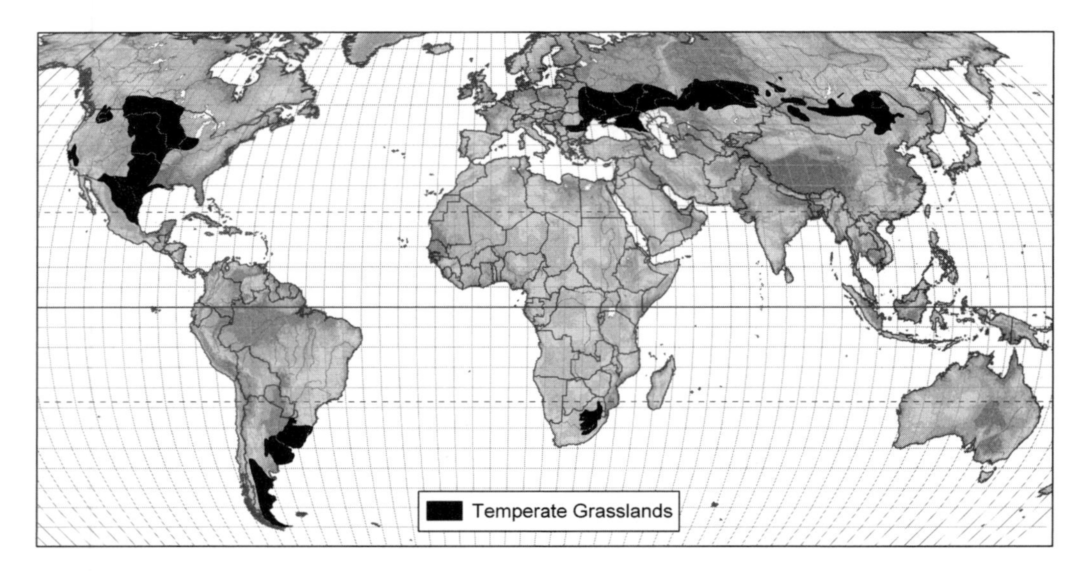

Figure 2.1 World distribution of the Temperate Grassland Biome. *(Map by Bernd Kuennecke.)*

Figure 2.2 Typical vegetation profile of a temperate grassland showing prevalence of forbs as well as grasses. *(Illustration by Jeff Dixon.)*

and replace each other in an east-west direction across the mid-continent. In other areas, latitudinal zonation is prominent and plant and animal communities change in a north-south direction. Still other natural grasslands are a consequence of rain-shadows on the lee sides of major mountain ranges or of high elevation, as is the case in southern Africa.

Temperate grasslands around the world have been so heavily altered by human activities (especially grazing of livestock, plowing under and conversion to agriculture, increased fire frequency, or cessation of grassfires) that little survives that is truly natural. Fire may have been the first way people managed grasslands. Annual burning eliminated tree and shrub seedlings and stimulated the growth of new grass shoots. Deliberate burning may have occurred as soon as fire could be managed to attract the wild grazing animals upon which hunting and gathering peoples, and indeed, early agriculturalists, depended for protein. Repeating firing of the forest edge may have expanded grasslands into wetter climate regions, such as the Prairie Peninsula of the United States and the eastern pampas of Argentina and Uruguay. Later, farmers and townsfolk stopped all wildfires, and trees invaded the grasslands.

Pastoralism—a nomadic way of life dependent on domestic livestock—in all likelihood first developed in the temperate grasslands of the Old World. Shifts in plant species abundance no doubt accompanied the spread of pastoral societies, since livestock select the most palatable plants. Weedy grasses and forbs evolved in response to the trampling and cropping of the vegetation by cattle, sheep, and other hoofed grazers. Tolerant of disturbance and preferring open habitat, weeds typically germinate rapidly in full sun, have short life spans, produce large numbers of easily dispersed seeds, and thus spread quickly to new disturbed sites. Not surprisingly, Old World weeds followed cattle, sheep, and goats when the animals were transported to other parts of the world.

What's in a Name?

In North America, French fur traders were the first Europeans to venture onto the grass-covered rolling hills in mid-continent that American pioneers would later call a sea of grass and cross in prairie "schooners." The French referred to it as a meadow or *prairie*. Today, the term prairie is often reserved for the tall-grass and mid-grass expressions of the biome.

The vast grasslands of Eurasia were named by their Russian and Ukrainian inhabitants. The word came into German as *steppe* and gained widespread use in the western European languages. In the United States, the term steppe often designates the dry short-grass prairie.

Guarani people lived in the fertile grass-covered lowlands of South America between the Uruguay and lower Paraguay rivers. Their word for "level plain" (*pampas*) was adopted by Spanish-speaking colonizers to refer to the temperate grasslands in the southern regions of that continent. Portuguese settlers in southeastern Brazil called the same vegetation *campos*.

In South Africa, Afrikaaners, the descendents of Dutch colonists, named the upland grasslands *veld*, a word that is narrowly defined as "field" but that implies a wildness or remoteness, a sense of the "outback." Plant geographers today use the term broadly with modifying prefixes to refer to a variety of vegetation types: highveld, grassveld, bushveld, and so forth.

The soils beneath prairies, steppes, pampas, and highveld turned out to be among the most fertile on Earth. Once the technology was available to break through the thick sod, many grasslands were converted to agriculture. Wheat and corn replaced native grasses and the temperate grasslands became the breadbaskets of their respective countries.

As people try to restore temperate grasslands today, they have come realize that a delicate balance must be achieved between protection and disturbance if native plants are to regain lost territory. Fire, grazing, and mowing all help to reduce woody plants and maintain a rich variety of herbaceous plants as well as the wild animals that live with them.

Geographic Location

Temperate grasslands occur in the middle latitudes (30° to 60° N and S) of both hemispheres. Specific latitudes depend on the continent on which they are located (see Figure 2.1). The largest areas of the Temperate Grassland Biome are found in the Northern Hemisphere on the North American and Eurasian continents. Smaller but still major segments of the biome are found in the Southern Hemisphere in South America and southern Africa.

North American and Eurasian grasslands owe their existence to their positions in the middle of large continents, where precipitation is reduced and temperature extremes are pronounced. The South American grasslands lie in the rainshadow east of the Andes Mountains. In Africa, elevation is a factor. According to latitude alone, these African grasslands are subtropical and would not qualify for the biome, but the elevated surface at the southern end of the African Plateau creates regional temperature patterns similar to middle-latitude conditions.

Australia has some subtropical grasslands that might qualify as part of the biome, but they are relatively minor on a global scale and are not discussed in this chapter. So, too, smaller areas of natural grassland such as the Texas's blackland and coastal prairies and Florida's flooded Everglades are not included.

Climate

Temperate grasslands have developed in regions having a semiarid climate (mostly BSk in the Koeppen climate classification) (see Figure 2.3). Mid-latitude semiarid climate regions receive 10–20 in (250–750 mm) of precipitation a year. In some areas (most notably the mid-continental prairie of North America and the East European steppes of Eurasia), a significant amount falls as snow during the winter months, such that snowmelt contributes importantly to soil moisture in early spring. The interior position of grasslands on the large North American and Eurasia continents results in the wide annual range in temperatures, a temperature pattern that is considered typical for the biome. Winters see temperatures well below freezing, whereas summers can be scorching. Such a temperature pattern is described by climatologists as continental. Similar temperature conditions often occur in humid forested regions at the same latitudes.

Seasonal temperature differences are first and foremost a consequence of the latitudinal location of the biome. Daylength (photoperiod) and the angle at which sunlight strikes the Earth's surface varies so much from one time of year to the next that surface temperatures are affected. Large landmasses react especially strongly to differences in the amount and intensity of incoming solar energy. Unlike the seas, continents heat up rapidly during the long days of high-angle sun in the

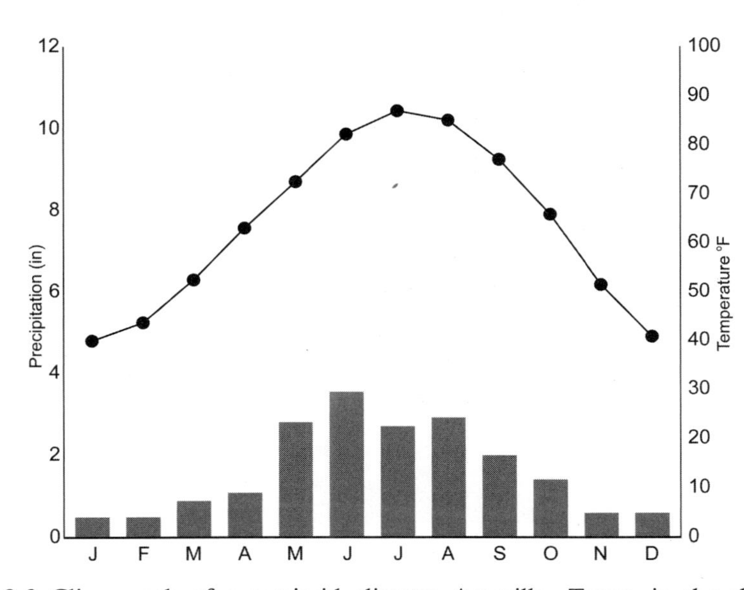

Figure 2.3 Climograph of a semiarid climate: Amarillo, Texas, in the short-grass prairie. *(Illustration by Jeff Dixon.)*

summer and then, during the shorter days and lower-angle sun of the fall and winter lose heat rapidly. The result is a great range of temperatures over the course of a year at middle and high latitudes, a pattern known as continentality. The effect is lessened the closer a place is to an ocean, where the moderating influence of the sea on temperatures is felt. Therefore, extreme temperatures do not occur in the grassland regions of South America's narrow southern tip.

In eastern Eurasia, the great annual range in temperature is responsible for the establishment of the Asian monsoon system. The very cold winters cause high atmospheric pressure to dominate the landmass at that time of year and generate dry air masses that move out of Asia. The reverse happens in summer, when low atmospheric pressure dominates the hot Asian continent and draws moist air into the continent from over the Pacific and Indian oceans. Much of the moisture is lost before the air masses ever reach the interior, so semiarid (and arid) conditions develop. In China, the wetter parts of the steppes are in the east and the drier parts are in the west at the greatest distance from the sea.

In North America, the precipitation pattern is affected by seasonal shifts in the Polar Front, the contact zone between colder polar and warmer subtropical air masses. The warmer air is less dense and is forced to rise over the colder air when they meet, generating precipitation. In the winter, the Polar Front migrates into lower latitudes and brings snow to the Great Plains and Central Lowlands. In the summer, it shifts poleward and draws warm, moist subtropical air into the continent's interior. Heated by the land below, this air can rise and produce rainfall. Because warmer air can hold more water vapor than cooler air, more moisture is

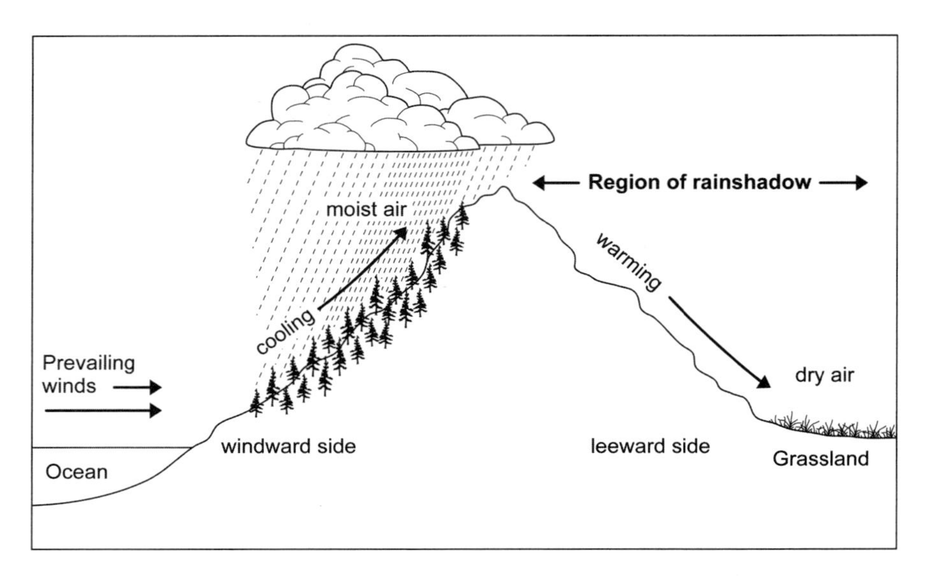

Figure 2.4 The rainshadow effect is a consequence of the downslope movement air on the leeward side of mountain ranges. As the air mass descends, it compresses and warms, reducing the relative humidity. This process causes the semiarid climates east of the Rocky Mountains, which are associated with short-grass prairies, and east of the Andes, where the Patagonian steppe occurs. *(Illustration by Jeff Dixon.)*

available in the summer and peak precipitation occurs in June and July. However, the distance from the oceanic sources of moist air means that much of the moisture is lost and not replenished before the air reaches the interior. An east-west component to the pattern arises because of the position of the Gulf of Mexico, a major source of summer moisture. Eastern prairies receive more rainfall than the western prairies, which are farther inland. Indeed, the Prairie Peninsula that extends into Illinois and Indiana does not have a semiarid climate at all, but rather one that is classified as subhumid.

Another major factor creating an east-west gradient in moisture in North American and, indeed, South American temperate grasslands, is the rainshadow effect (see Figure 2.4). Both regions lie in a zone of prevailing westerly winds, and the storms that develop along the Polar Fronts of both hemispheres move from west to east. In both instances, as the air masses pass across the continents, they must rise over major mountain ranges—the Cascades, Sierra Nevada, and Rocky Mountains in North America and the Andes Mountains in South America. The rising air cools and releases much of its moisture on the west-facing slopes. After the air masses cross the mountains, they descend the eastern slopes. The downward flow causes air temperature to increase again because of a process known as adiabatic warming. Warming makes the air absorb moisture rather than release it, so precipitation is greatly reduced on the lee side of the mountains. The resulting dryness is known as the rainshadow effect.

In all regions of the biome, gradual changes in climate occur across great distances. In Eurasia, the variation occurs in a north-south direction. The wetter (subhumid) northern regions of the Temperate Grassland Biome border the Temperate Broadleaf Deciduous Forest and Boreal Forest Biomes and the drier (arid) southern margins grade into the Desert Biome. In the Americas, the gradation from subhumid to arid runs in an east-west direction, with precipitation becoming progressively less the farther west one goes. In all areas, a change in vegetation coincides with the change in total precipitation. Wetter areas contain taller grass species and usually a greater variety of plants and animals than do the shorter grass areas in the driest parts of the biome.

Vegetation

Perennial grasses are the most common growthform in this biome; however, a large number of other herbaceous plants, primarily perennial forbs, are also found (see Figure 2.5). Particularly abundant are species in the sunflower and pea (or legume) families. Among the grasses, both sod-forming and bunchgrasses (see Chapter 1) are prominent throughout, but bunchgrasses become a much more frequent component of the vegetation in the drier short-grass regions.

Woody plants are not major components of the vegetation, except in the driest regions or along waterways or in disturbed areas. Shrub invasion is a key indicator of degraded grasslands, usually the result of overgrazing by domestic livestock.

Figure 2.5 Konza Prairie in the Flint Hills of Kansas. *(Photo by Edwin Olson.)*

Grasses and forbs grow to different heights at maturity. Furthermore, some grasses and most forbs are erect, while others are recumbent, creeping along the ground. Thus, a distinct layering of plant foliage occurs in some temperate grasslands. Forbs and grasses may grow and bloom at different times during the growing season, so temperate grasslands may have recognizable color phases, or aspects, depending on what is blooming. This is especially evident in the east European steppes.

Grasses and perennial forbs are well suited to withstand cold seasons and tolerate grazing and fire because of the position of their renewal buds near the ground surface. The plants are generally able to regenerate from these buds after seasonal diedowns, burns, or cropping by grazing animals.

Soils

Formation of temperate grassland soils. Chapter 1 addressed how the aboveground portions of the plant adapted grasses to the challenges of their environment. The underground portion of the grass plant is equally, if not more, important to the functioning of the ecosystem and formation of soil. Grasses develop a dense, intricate system of fine roots that not only acts to hold the plant in place but also absorbs water and nutrients from the soil and stores some of the carbohydrates that the plant manufactures during photosynthesis for later use. In fact, more living plant matter (biomass) exists below ground in the temperate grasslands than above. Roots continually grow, die, and decay and thus contribute organic matter in the form of humus to the soil. The seasonal die-off of the aboveground parts of the grasses also adds large amounts of humus to grassland soils. Humus attracts some ions that are important plant nutrients and prevents them from being removed from the root zone when rainwater percolates down through the soil. The fine roots create a crumbly texture to the soil that allows water and air to penetrate to depths up to 12 ft (3.5 m) in the wetter parts of the biome, further enhancing the soil environment for plant growth. Those soluble compounds (for example, calcium carbonate and magnesium carbonate) that might be leached to depth after rainstorms or snowmelt can be returned to the root zone during dry periods by capillary action along the tiny tunnels left by dead roots. The carbonates precipitate out near the top of the subsoil (B horizon) when the soil moisture evaporates. The presence of a concentration of these bases can be determined in the field by chemical tests. Sometimes, they form white nodules visible to the naked eye. The concentration of carbonates in this manner is the main feature of the soil-forming process known as calcification, which is characteristic of semiarid and arid climate regions. The carbonate-rich layer forms closer to the surface the drier the climatic conditions (see Figure 2.6).

Many grasses have symbiotic relationships with nitrogen-fixing bacteria that grow in visible nodules along the roots. These bacteria convert pure nitrogen in the soil gases to nitrogen compounds such as nitrates that can dissolve in the soil water

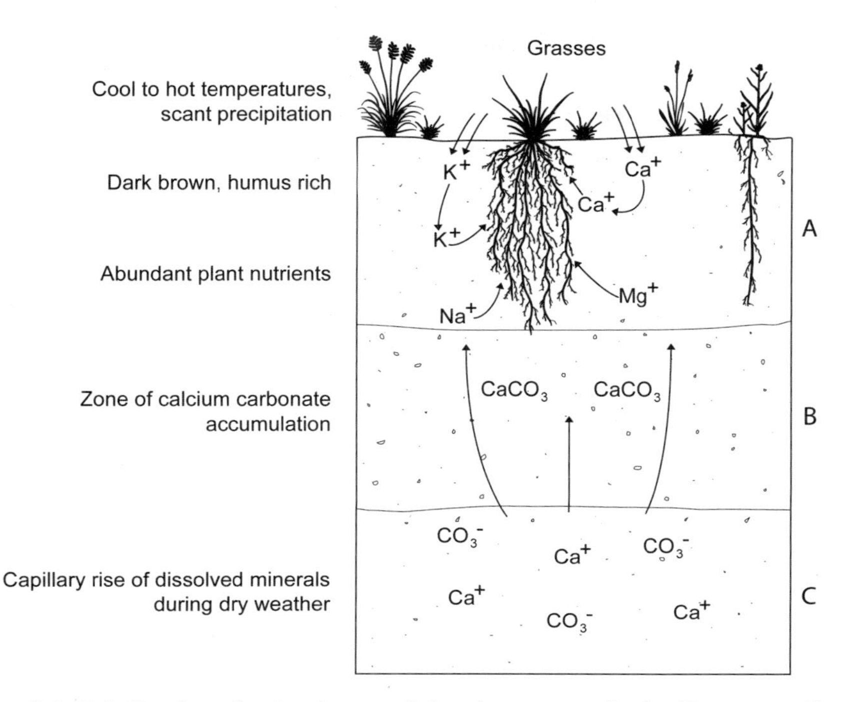

Figure 2.6 Calcification, the dominant soil-forming process in the Temperate Grassland Biome, conserves ions of calcium (Ca^{++}) and other plant nutrients in the soil column as evaporation draws soil moisture containing dissolved minerals up toward the surface. *(Illustration by Jeff Dixon.)*

and be taken up by the plants. Nitrogen is often a limiting factor for plant growth in terrestrial ecosystems, so this relationship between grasses and bacteria is just one more way that grasses, in combination with a semiarid continental climate, produce some of the most fertile soils on Earth.

Soil types. Grassland soils of temperate regions are classified in U.S. Soil Taxonomy as mollisols. The prefix "moll" means soft and describes the friable or crumbly texture of these soils that persists even when the top layers dry out. (Other parts of the world and the United Nations Food and Agriculture Organization use other names. In Canada, they are called chernozemic soils; in Russia they are chernozems.) Mollisols are brown in color because of the abundance of humus in both the topsoil and subsoil (A and B horizons) (see Plate I).

Color varies from dark brown and even black to chestnut brown and light brown as climate changes from subhumid to dry and grasses change from tall with deep roots to short with shallow roots. For the darkest of all, the truly black soils, the Russian name *Chernozem* (meaning black soil) is used, even in the United States. Chernozems have especially deep A and B horizons (each up to 3 ft or 1 m deep), both of which are dark with humus and humic acid stains. They have formed where semiarid regions receive the most precipitation and where the parent material is a fine, windblown material known as loess. Loess originated as tiny particles on the newly exposed bare ground at the edge of continental glaciers when ice sheets retreated at the end of the Pleistocene ice age. Winds coming off the ice picked the dust-like material up and carried it to more humid, vegetated regions, where it accumulated. Thick deposits of loess occur in the Central Lowlands of North America, across much of semiarid Eurasia, and in Uruguay and eastern parts of Argentina. Loess is often rich in carbonates. The soils developed on it are among the most fertile in the world, and today they produce much of the world's wheat and corn (both of which are domesticated grasses).

Animals of Temperate Grasslands

The herbivorous mammals of the temperate grasslands generally can be placed in one of two groups: fast-running ungulates that move in either small or large herds and burrowing rodents that live in colonies. It appears to be an advantage in an open habitat to have many sets of ears and eyes alert to danger. The largest predators, at least in North America and Eurasia, were wolves, but they have been largely exterminated. The smaller predators surviving today include members of the dog, cat, and weasel families.

Animal diversity is relatively low, especially in comparison with the animals of some tropical grasslands. Large mammals are generally rare or already extinct. Different species are found in each part of the biome. Consult the appendix to this chapter for details.

By necessity, birds in the grasslands are ground-nesters and some are good runners. Partridge- or grouse-like birds are common. Reptiles are represented by both snakes and lizards, but neither group is especially diverse. Amphibians are primarily the tailless forms, the frogs and toads, and are associated with ponds and streams.

By far the most important herbivores are insects. Grasshoppers are especially abundant.

Regional Expressions of the Temperate Grasslands Biome

The North American Prairies

Several areas of natural grassland are found on the North American continent (see Figure 2.7). By far the largest continuous grassland stretches from the foot of the Rocky Mountains east to approximately the 100th meridian in the middle of the continent. It covers much of the landform region known as the Central Lowlands. The so-called Prairie Peninsula extends eastward from this major belt of prairie across Iowa and northern Missouri into northern Illinois and Indiana.

Over this vast area, three types of prairie may be distinguished, each with different typical plants and animals. From east to west are the tall-grass prairie, the

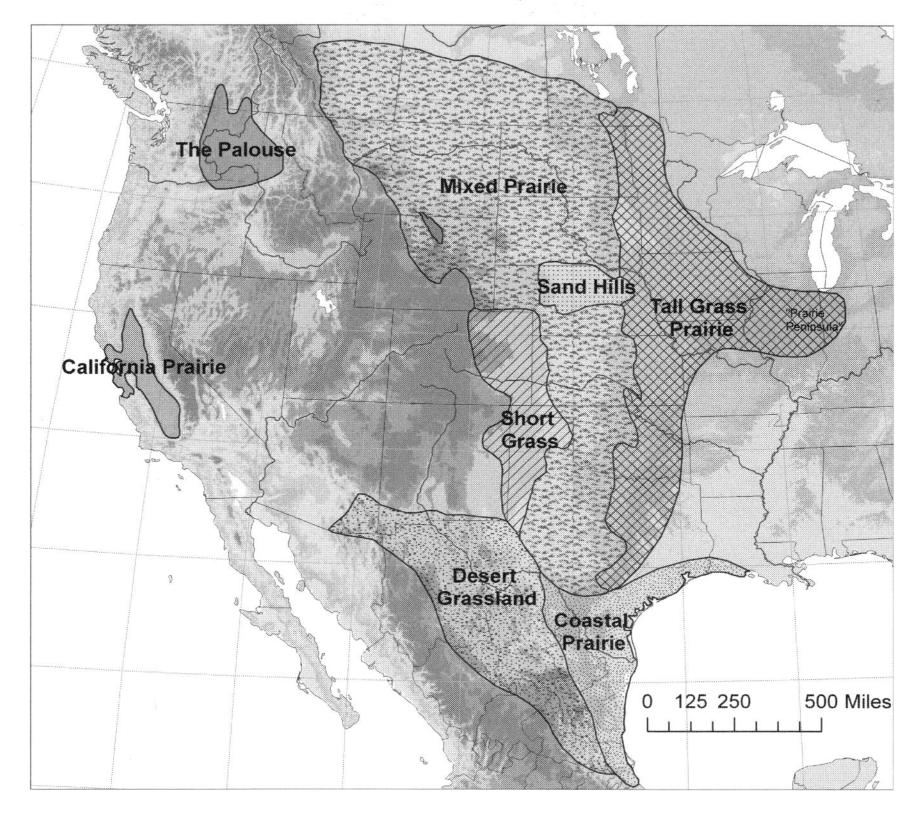

Figure 2.7 Distribution of the North American prairies. *(Map by Bernd Kuennecke.)*

mixed prairie, and the short-grass prairie. Smaller segments of the biome exist west of the Rocky Mountains, including the Palouse prairies that occur from eastern Washington state and Oregon through parts of Idaho into northwestern Montana, the California grasslands in that state's Central Valley, and desert grasslands found from southeastern Arizona to western Texas and south into the Mexican states of Sonora, Chihuahua, and Coahuila.

Tall-Grass Prairie

Tall-grass prairie is characterized by perennial C_4 grasses that reach heights of 6 ft (2–3 m) or more when they mature in late summer. It occupies the eastern, subhumid part of the biome that occurs in a band from southern Manitoba, Canada, south to Gulf Coast of Texas and includes the Prairie Peninsula, where annual precipitation is considerably more than 20 in a year.

The combination of repeated burning of the grasslands to the west by prehistoric peoples and the grazing of the bison attracted by the new flush of grass that followed the fires are believed to have expanded the prairie into areas where, without such disturbances, forest might have flourished. The ability of grasses to resprout from renewal buds close to the ground would have given them an advantage over trees whose buds, held high above the ground on seedlings as well as on older specimens, would have been destroyed. Today, remaining areas of tall-grass prairie are maintained by prescribed burns and the grazing of both wild (bison) and domesticated (cattle) animals.

Although total annual precipitation is higher than expected for the Temperate Grassland Biome, most is received in the summer months as typically happens in the drier parts of the mid-continent. The temperature pattern of very cold winters and warm to hot summers is definitely consistent with the tall-grass prairie's position in the interior of the continent and typical of mid-latitude grasslands in general.

Tall-grass prairie grows on unconsolidated parent materials derived from glacial deposits and loess dating back to the Pleistocene. The resulting mollisols have deep horizons and contain much humus from decaying plant material that gives them a dark blackish brown coloration. The deep roots of the grass and capillary action bring calcium carbonate up into the subsoil (B horizon) and raise the pH to neutral or slightly basic levels. Distinct carbonate nodules do not usually form. True chernozems have formed in this part of the biome. A major exception to the characteristically fertile soils occurs in the Flint Hills in eastern Kansas, where thin rocky soils cover limestone. Because of poorer soils, Flint Hill prairies support fewer plant species than areas on the richer mollisols.

Tall-grass prairie has the greatest number of native plant species of any of the mid-continent North American prairies. Some 265 species have been recorded from prairie in Iowa. Somewhat smaller totals have been reported from other areas. The very tall grasses are accompanied by numerous shorter grasses and many perennial forbs. On close inspection, the prairie may look more like a meadow of wildflowers than a grassland (see Plate II). Big bluestem (see Figure 2.8) is the characteristic tall sod-forming grass, little bluestem is a common mid-size bunchgrass, and sideoats grama is an important short bunchgrass. Many prairie forbs, such as yarrows, coneflowers, and various sunflowers, have been taken into cultivation and have become popular garden plants.

The tall-grass prairie was once the "home on the range" where the "buffalo roam and the deer and the antelope (that is, pronghorn) play," but these large mammals and elk are now gone. Also absent is the wolf, which once was the main large predator. Today, coyotes and red foxes have taken over as the main carnivores of the eastern grasslands. Mammalian herbivores include rabbits and hares and burrowing rodents such as ground squirrels and mice. The pocket gopher, member of a family of rodents found only in North America, is also associated with this part of the biome. Numerous kinds of songbirds, many of them seed-eating sparrows, nest in the tall-grass prairie. The widespread Burrowing Owl occupies abandoned rodent burrows. The largest ground bird, the Greater Prairie Chicken, is now rare. An introduced species, the Chinese Ring-necked Pheasant, has more or less replaced it.

Figure 2.8 Big bluestem, sod-forming grass of the tall grass prairie. *(USDA-NRCS PLANTS Database/Hitchcock, A. S. [rev. A. Chase]. 1950.* Manual of the grasses of the United States. *USDA Miscellaneous Publication No. 200. Washington, DC.)*

Tall-grass prairie may once have covered 400,000 mi^2 (150,000 km^2). Most prairie has long since vanished under the plow. The soils that made this biome so rich in wild plants and animals were well suited to agriculture once a steel plow was invented and the thick sod could be broken. The region became the Corn Belt of the United States and sustained hog farming and feedlot cattle raising. In the states of Illinois, Indiana, Iowa, and North Dakota and the province of Manitoba less than 1 percent of the pre-European settlement tall-grass prairie remains. Minnesota and Missouri, two states that have promoted conservation, have been able to preserve only about 9 percent of their grasslands. No large tracts survive; many reserves are less than 20 ac (0.08 km^2) in size. Important remnants occur in the Loess Hills of western Iowa, the Prairie Coteau in eastern South Dakota, the Flint Hills in Kansas and Oklahoma, Osage Hills in Kansas, and the Fort Worth Prairie in Oklahoma and Texas. In 1996, Tallgrass Prairie National Preserve was established in the Flint Hills, preserving 180 ac in one of the largest areas of intact tall-grass prairie in the United States. At Konza Prairie, also in the Flint Hills, The Nature Conservancy owns another 8,600 ac, which are managed by Kansas State University as a research natural area.

Pocket Gophers and Pronghorns: American Originals

Members of two of the four mammalian families endemic to North America are prominent elements of the prairie fauna: the pocket gophers (Geomydiae) and the pronghorn (Antilocapridae) (see Figure 2.9). One other, the pocket mice and kangaroo rats (Heteromyidae), is represented, but it is widespread and not so strongly associated with grasslands.

Figure 2.9 Pronghorn, the only member of a family of ungulates unique to North America and the continent's fastest mammal, on the winter prairie. *(Photo © Ed Endicott/Wysiwyg Foto LLc and Shutterstock.)*

Pocket gophers possess two fur-lined pouches on the sides of their head, outside their mouths. In them, they carry food to underground storage rooms, where they empty the pockets by turning them inside out. These small-to medium-size rodents are thickset and appear to have no necks. Tails are short. The massive skulls are flattened and well adapted to a fossorial life. Solitary, a pocket gopher spends most of its time below ground, excavating tunnels and the chambers it uses for nesting, storage, and bathrooms. Its lips actually close behind the incisors so it can loosen dirt and cut through roots with its teeth without getting dirt in its mouth. Most digging is done with its strong claws. Pocket gophers feed mostly on roots and tubers.

Pronghorns are America's fastest land animal. Maximum bursts of speed of 54 mph (86.5 km/hr) have been recorded for this beautiful ungulate, while small herds run across the open prairie at a steady 26 mph (42 km/hr). Pronghorns both browse and graze and, if necessary, can subsist on water derived from their diet. Taxonomically they are more closely related to deer than cattle. Thirteen genera are known from the Pliocene and Pleistocene. Today only one genus with a single species survives.

Some pocket mice and kangaroo rats are found in the drier prairies. Like pocket gophers, they have external pouches. However, their skulls are papery thin and jaws small and weak. They construct burrows under bushes and eat mostly seeds and insects.

Mixed-Grass Prairie

Species of the tall-grass prairies to the east mix with those of the short-grass prairie to the west across a vast stretch of territory in the middle of the North American continent. This is the mixed-grass prairie or simply mixed prairie. Beginning in the north at about 52° N latitude in southern Saskatchewan and neighboring parts of the adjoining prairie provinces of Canada, this part of the Temperate Grassland Biome extends some 23° of latitude southward across the Great Plains to Texas (see Figure 2.7). Elevations rise from 1,600 ft (500 m) on the eastern margins of the mixed-grass prairie to 3,600 ft (1,100 m) in the west, where it grades into short-grass prairie.

This region has a true semiarid climate. Precipitation totals range from about 24 in a year (600 mm) in the east to only about 14 in (350–400 mm) a year in the west. The continental temperature pattern of hot summers and very cold winters prevails.

Because of lower amounts of precipitation than what falls farther east, the soils here are not as deep, and decaying roots and tillers not as abundant as in the tall-grass prairie. Mollisols are typically brown or chestnut in color, rather than black. However, one of the largest intact samples of mixed-grass prairie today is not associated with mollisols at all, but rather is found in the Sand Hills of Nebraska on entisols weakly developed on fossil sand dunes. The grasses of Nebraska's Sand Hills are a unique assemblage dominated by sand bluestem and needle-and-thread grass.

Vegetation has two layers consisting of mid-size C_3 grasses that attain heights of 2–4 ft (60–120 cm) and short C_4 grasses that may be only 0.6 to 2 ft. (20–60 cm) high. Western wheatgrass (see Figure 2.10), dropseed, and needle-and-thread grass dominate and form the top layer of the vegetation. Blue grama and buffalograss are the main grasses of the lower layer.

Figure 2.10 Western wheatgrass, a common grass of the mixed prairie. *(USDA-NRCS PLANTS Database/USDA NRCS.* Wetland flora: Field office illustrated guide to plant species. *USDA Natural Resources Conservation Service.)*

Some tall C_4 grasses of the tall-grass prairie, such as big bluestem and Indiangrass, also occur. The mid-size grasses are dominant and are cool-season plants, meaning they sprout early in the spring and mature by early summer. The short grasses, on the other hand, are warm season growers and do not reach maturity until late summer. This combination of grass strategies means that green grass is available to grazing animals throughout the growing season and often into early December.

Perennial forbs are less common in mixed-grass prairie than in tall-grass prairie. Yarrows and asters, members of the sunflower family, are conspicuous elements of the forb community, as is plains beebalm, a mint. Some woody plants appear in this part of the biome both as subshrubs (for example, fringed sagebrush) and dwarf shrubs (for example, wild roses). In the prairie provinces of Canada, trees (mostly trembling aspen) invaded the grasslands when prairie fires were prevented, so that today much of the landscape is actually forested. Some succulents also make their appearance in this part of the Temperate Grassland Biome. These are generally well-armed species of prickly pear cactus.

Animals of the mixed-grass prairie are quite similar to those of the tall-grass prairie, although a greater variety of reptiles, including rattlesnakes and horned lizards, is associated with this part of the Temperate Grassland Biome, reflecting the drier conditions of the region. Large native mammals have, for the most part, been exterminated, and the prairie dog is a keystone species. Without its activities, the vegetation and animals of mixed prairies (and short-grass prairies) would be noticeably different. Prairie dogs are actually squirrels. Early settlers thought their alarm calls sounded like barks, so they called these rodents dogs. Five species (see Table 2.1) differ from each other in tail length, presence or absence of a dark line above the eye, and vocalizations, and occur in different geographic locations. Most widespread is the black-tailed prairie dog, whose range corresponds closely with that of the mixed prairie.

Black-tailed prairie dogs are colonial animals that live in burrows (see Figure 2.11). A complex of several colonies or "towns" may cover an area more than 60 miles (1000 km) wide. Individual prairie dogs stand about 12 in (30 cm) tall and weigh about 1.5 lbs (700 g). They are active by day, unlike most other small mammals in the

Table 2.1 Prairie Dog Species

LONG-TAILED GROUP	
Black-tailed prairie dog	*Cynomys ludovicianus*
Mexican prairie dog	*Cynomys mexicanus*[a]
SHORT-TAILED GROUP	
Gunnison's prairie dog	*Cynomys gunnisoni*[b]
White-tailed prairie dog	*Cynomys leucurus*[b]
Utah prairie dog	*Cynomys parvidens*[b,c]

Notes: [a]Federally listed as endangered; [b]This species hibernates; [c]Federally listed as threatened.

grasslands, so they are important prey for a number of daytime predators, including American badger, coyote, Golden Eagles, and Ferruginous Hawks. Because they do not hibernate, they remain a food source for carnivores throughout the year.

Most prairie dog predators eat a wide variety of prey. Only the very rare and endangered black-footed ferret (see Plate III) eats prairie dogs exclusively. It is also the only nocturnal predator of prairie dogs, seeking them out as they sleep in their burrows. It takes a very large colony of prairie dogs to sustain a population of ferrets.

Openings to prairie dog burrows are surrounded by mounds that may be 2.5 ft (0.75 m) high and up to 7 ft (2 m) in diameter. The dirt crater ventilates underground passages and prevents their flooding. The mounds also serve as lookout points on which members of the colony can stand watch. By digging burrows and bringing top-soil and subsoil up from below, prairie dogs retrieve nutrients washed downward in the soil and deposit them on the surface, enriching the mound and its immediate vicinity. Some plants are essentially found only on the mounds (see Table 2.2). The grasses and forbs that grow there tend to have faster growth rates and higher protein content than similar plants growing at some distance from mounds. The mound plants are more easily digested by large herbivores than plants elsewhere on the prairies. Such high-quality forage attracts pronghorn, bison, and domestic cattle into the colony's territory.

Figure 2.11 Black-tailed prairie dog, keystone species in the mixed-grass prairie. *(Photo by John and Karen Hollingsworth, U.S. Fish and Wildlife Service.)*

Table 2.2 Plants Nearly Restricted to Prairie Dog Mounds

Black nightshade	*Solanum americanum*
Fetid marigold or prairie dog weed	*Dyossodia papposa*
Pigweed	*Amaranthus retroflexus*
Scarlet globemallow	*Spaeralcea coccinea*

War on Dogs

Despite the usefulness of prairie dogs in maintaining a healthy natural ecosystem, their numbers have been purposefully and drastically reduced. Prairie dog holes were leg-breaking hazards for ranchers' horses and crop-reducing nuisances for farmers. Widespread poisoning began in the early 1800s. Where they survived, prairie dogs came to represent poor land stewardship. By the early 1900s, the federal government launched campaigns to eradicate the rodents from the central grasslands of North America. Two hundred years ago some 5 billion black-tailed prairie dogs may have lived on the 395 million ac (160 million ha) of prairie between southern Saskatchewan and southern Texas. Today, only 2 percent of the former range is occupied, most of it east of the 102° W meridian. Recurring outbreaks of sylvatic plague—a fatal disease caused by the bacterium *Yersinia pestis*, which was probably introduced from China in the early twentieth century—decimated populations in the western two-thirds of the former range. There were too few dogs to support black-footed ferrets and they faced imminent extinction. Conservations efforts are under way, but government-directed and privately initiated poisoning of colonies continues in other places. Recreational shooting and habitat loss still contribute to losses. Restoration of large prairie dog colonies would conserve a variety of prairie plants and rare animals, such as ferrets, Mountain Plovers, and Burrowing Owls, and improve some of our nation's natural grazing lands.

Black-tailed prairie dogs, too, graze on the colony grounds, and this activity keeps the vegetation about 3–4 in (8–10 cm) high. Furthermore, presumably to increase visibility and the likelihood of spotting approaching predators, they clip plants that they do not eat to heights of less than 12 in (30 cm). This not only helps prevent the invasion of woody plants such as mesquite, but also creates favorable nesting habitat for Mountain Plover and Horned Lark. Burrowing Owls (which do not eat prairie dogs), badgers, prairie rattlesnakes, and even tiger salamanders use abandoned and active burrows for nesting dens or shelter. The clipped grass in a large prairie dog "town" may also serve as a firebreak, slowing and altering the passage of wildfires.

Much of the mixed-grass prairie has been converted to the production of wheat (spring wheat in the north; winter wheat in the south) and oil seeds (sunflower and rapeseed or canola). The drier parts of the regions are used as grazing land for open-range cattle. Heavy grazing pressures have caused a decline in mid-size grasses, and therefore the proportion of short grasses on rangeland is larger than on ungrazed lands. Aspen has invaded vast stretches of the prairie land of Canada since the cessation of wild fires. Although the destruction of the mixed-grass prairie has not been as complete as that of the tall-grass prairies, most states and provinces have seen a 60–80 percent decline in the area covered since the beginning of European settlement. The Sand Hills of Nebraska and adjacent areas of South Dakota represent the largest expanse of native grassland remaining in the United States. Like the Flint Hills in the tall-grass prairie, these grasslands were saved in part because their soils were unsuited for cultivation.

Short-Grass Prairie

The short-grass prairie, the westernmost part of the North American prairies, is confined to the central Great Plains from southeastern Wyoming through eastern Colorado and New Mexico to the Texas and Oklahoma panhandles (from roughly 41° N to 32° N latitude). The driest of the mid-continent prairies, short-grass prairie covers some 175,000 mi^2 (280,000 km^2). Precipitation totals range from 12–22 in (330–550 mm) a year. Although the latitude is subtropical, the interior location on the continent subjects the region to a wide range of temperatures annually.

With lower precipitation, soil development is not as deep as in either the tall-grass or mixed-grass prairies. The humus-bearing horizon is quite shallow and white calcium carbonate nodules appear some 9 or 10 in (25 cm) below the surface. Soils tend to be light brown in color.

The vegetation is dominated by two grasses, blue grama and buffalograss (see Figure 2.12). Both are shallow-rooted, warm season species and both stand 4–16 in (10–40 cm) tall.

Figure 2.12 Buffalograss, a dominant grass of the short-grass prairie. *(USDA-NRCS PLANTS Database/Hitchcock, A. S. [rev. A. Chase]. 1950.* Manual of the grasses of the United States. *USDA Miscellaneous Publication No. 200. Washington, DC.)*

An open layer of mid-size grasses, including needle-and-thread grass, western wheatgrass, and wire grass rises above the short-grasses in areas that have not been grazed. Perennial forbs are essentially absent, but woody shrubs and subshrubs such as some sagebrushes are present, as are prickly pear cactus. The abundance of woody plants and succulents becomes greater the more heavily the area has been grazed.

Great herds of bison once passed through the short-grass prairies on periodic migrations in search of better forage (see Figure 2.13). The movements of today's remnant herds are restricted and carefully managed. Now the major mammalian herbivores are jackrabbits and prairie dogs.

Two small burrowing mammals are worth noting, because they are members of a family endemic to North America. These are the plains pocket gopher and northern pocket gopher.

Sparrows and finches are abundant. Some, such as the Lark Bunting and Brewer's Sparrow, have ranges extending into the deserts of the American Southwest. Horned Larks and Western Meadowlarks are common. The Mountain Plover replaces the Upland Sandpiper of moister prairies as a shorebird nesting on the grasslands.

Reptiles in general are well adapted to dry conditions. Snakes such as prairie rattlesnakes, hog-nosed snakes, and gopher snakes are abundant. There are also skinks, a horned lizard, and a box turtle.

Figure 2.13 Short-grass prairie with bison, Custer State Park, South Dakota. *(Photo © Jim Parkin / Shutterstock.)*

The Mammoth-Steppe: What Might Have Been?

During the last great ice age treeless vegetation stretched from Eurasia across the Bering Land Bridge into North America. Described as either steppe-tundra or mammoth-steppe, the biome is now extinct. Several possible explanations have been put forth regarding its demise, but no consensus has been reached.

Pollen data, the fossil record of animals present, and even food particles stuck in the teeth of mummified mammoths suggest that a nutritious mix of sedges, short bunchgrasses, forbs, mosses, lichens, and shrubs once covered a large portion of the northernmost lands of the Northern Hemisphere. Mammoths, mastodons, giant bison, large muskoxen, caribou, large sheep, camels, horses, and saiga antelopes foraged across the region. Large carnivores such as dire wolves, cave bears, lions, and cheetah preyed upon the megaherbivores. Now they are all gone.

Some scientists see clear evidence that the rapid warming at the end of the Pleistocene changed the vegetation, but was too fast to allow large, slowly reproducing mammals to adapt. Another interpretation is that the animals disappeared before the steppe vegetation, and the loss of their cropping, manuring, and other activities spelled the end for the biome. Still others suggest that the Bering Land Bridge allowed skilled hunters from eastern Siberia to enter North America. Hunting increased death rates among the largest animals to levels higher than birth rates. Populations could not maintain their numbers and were doomed.

The loss of the mammoth-steppe biome likely was due to the combination of such factors. But *if* people were largely or even partially responsible for the extinction of the largest land mammals on the North American continent, this represents a major human impact on Earth. And if it had not happened, mammoths, giant bison, dire wolves, and lions might still be seen at home on the prairies. Can temperate grasslands be restored by introducing relatives of these long gone creatures that survive in other parts of the world? Some think so.

The short-grass prairie has not suffered as much outright destruction as have the tall-grass and mixed-grass parts of the biome, but impacts from human use have definitely altered the composition of the vegetation. Much of the area is rangeland and much has been overgrazed. As a result, the mid-size grasses have disappeared and thorny or otherwise unpalatable perennial and annual forbs and shrubs have invaded. An abundance of prickly pear cactus is an obvious indicator of overgrazing. Some short-grass prairie is under public ownership on the National Grasslands, administered by the U.S. Forest Service, but these, too, were severely degraded by overgrazing. The short-grass prairie is actually expanding northward as a consequence of damage to the mixed-grass prairie related to overgrazing.

Parts of the short-grass prairie were plowed up for the irrigated cultivation of alfalfa, corn, sugar beet, and cotton. Much land has been put into the production of wheat using dry-farming techniques.

Other Major North American Temperate Grasslands

The Palouse Prairies

A rich grassland once existed in the Intermountain West on the Columbia Plateau between the Cascade Mountains and the Rocky Mountains. The main section of vegetation covered unglaciated parts of eastern Washington, northeastern Oregon, and northwest Idaho. Separated from this area by some distance is an outlier in northwestern Montana. The Palouse lies not only in the rainshadow of the Cascades, but it is also part of a region characterized by winter rains and summer drought. Dry summers, combined with the barrier to plant and animal movement imposed by the Rocky Mountains, have resulted in plant and animal life in the Palouse that differs from that of the mid-continent prairies.

In the Palouse, deep, fertile soils developed on loess that had been deposited on a bedrock of basalt. Soils were further enriched by volcanic ash from the explosion about 6,500 years ago of former Mount Mazama, the remnants of which today encircle Crater Lake in Oregon.

Native Palouse vegetation was dominated by bunchgrasses, in particular bluebunch wheatgrass and Idaho fescue. A rich assortment of perennial forbs grew with the grasses. Cama, or wild hyacinth, grew—often in large numbers—in low-lying areas that held water in early spring and its bulbs were important food for Native Americans. Dwarf shrubs such as snowberry and wild roses were common, as were some small annual forbs.

The perennial grasses and forbs of the Palouse typically are green throughout the relatively mild winters and flower between May and September. Woody plants may have green foliage only in summer. The annual forbs typically germinate in midwinter, bloom in early spring, and then die back. They spend most of the year as part of the soil seed bank.

Plants of the Palouse are well adapted to frequent burning. When fires occurred often, relatively little dead plant material (that is, fuel) accumulated and the fires burned at low temperatures. After European settlement, prairie fires were quickly extinguished or prevented. This allowed for a buildup of fuel over many years and meant that, when the prairie did burn, the fires were hot and destructive. Many of the native perennial plants were killed off.

The animals of the Palouse are similar to those of the mid-continent prairies of North America. Close to the Rockies in Montana, grizzly bear and bighorn sheep would have foraged at lower elevations.

The Palouse has largely been destroyed by agriculture (mainly wheat production) and overgrazing. Today, it is highly fragmented into small remnants and suffers from the invasion of exotic annual grasses, especially cheatgrass (see Figure 2.14). Exotic annual forbs such as yellow star-thistle (*Centaurea solstitialis*) have more recently become serious problems.

An increase in native sagebrushes also occurs on overgrazed rangelands. Large areas of relatively intact Palouse prairie can be found in Hell's Canyon (in Oregon

Figure 2.14 Cheatgrass or downy brome, a species native to the Old Word, now covers large areas of the Palouse. *(USDA-NRCS PLANTS Database/Hitchcock, A. S. [rev. A. Chase]. 1950.* Manual of the grasses of the United States. *USDA Miscellaneous Publication No. 200. Washington, DC.)*

and Idaho) and Franklin D. Roosevelt Lake (in Washington state) national recreation areas and on the National Bison Range north of Missoula, Montana.

The California Prairie

The California prairie, like the Palouse prairie, has largely disappeared. It once covered about 25 million ac (10 million ha) in the foothills of the Central Valley and in some coastal valleys. The Central Valley lies between the coastal ranges of California and the Sierra Nevada. Its southern boundary is marked by the Tehachapi Range. This is another rainshadow region that has a mediterranean precipitation pattern of winter rain and summer drought. The southern half, the San Joaquin Valley, is considerably drier than the northern Sacramento Valley.

The original vegetation consisted of perennial cool-season (C_3) bunchgrasses. Purple needlegrass and Malpais bluegrass were dominant. A number of annual forbs and grasses grew between the clumps of grass, as did perennial bulbs such as

A Grass That Cheats

Cheatgrass (*Bromus tectorum*) is a problem. Native to Mediterranean Europe, North Africa, and Southwest Asia, it is an Old World weed well adapted to invade disturbed grasslands—something it has done across North America, Eurasia, South Africa, Australia, and even Greenland. Cheatgrass currently dominates more than 100 million ac in the Intermontane West.

Also called downy brome, drooping brome, military grass, and broncograss, cheatgrass got its most popular name from late-nineteenth-century wheat farmers who believed that their seed grain was contaminated with weed seeds. In other words, they had been cheated, and their fields began to produce more and more brome and less and less wheat. The new weed cheated against the native grasses of the western United States in other ways. Cheatgrass is a winter annual (as is winter wheat) and germinates in the fall. Seedlings go dormant during winter, but grow vigorously in early spring, getting a head start on the yearly regrowth of prairie perennials. For a short time early in the growing season, it provides nutritious forage for wildlife and livestock, but as the season progresses it cheats them out of fodder, because the mature spikelets are sharp and bristly and injure their eyes, ears, and mouths. By midsummer, little forage is available for cattle or wild sheep, deer, and elk.

Cheatgrass cures by late summer, when its dry stalks fuel grass fires. The fire cycle on the western plains has been reduced in some areas from as long as 60–100 years to a mere 10. Fires are now too frequent for native plants to withstand, but provide just the constant type of disturbance cheatgrass thrives on.

brodiaeas. In those years with substantial rain flowering forbs such as the bright orange California poppy (see Plate IV) and purple owlsclover carpeted the foothills.

The composition of the California grasslands changed after the arrival of Spanish settlers around 1775. This marked an introduction of cattle, sheep, and goats. Native perennials were not adapted to heavy grazing and died out. Annual plants, most of which were introduced from the Mediterranean parts of Europe, replaced them.

A number of the animals of the original California grasslands are endemic to the region at either the species or subspecies level. Examples include the tule elk, two endemic subspecies of kangaroo rat, the insect-eating San Joaquin antelope squirrel, the San Joaquin pocket mouse, and the San Joaquin Valley kit fox. In addition, there once were more wide-ranging mammals such as mule deer, grizzly bears, and wolves.

Today, it is estimated that less than 1 percent of grass species on remnant patches of California prairie are native species. The demise of the grasslands is consequence of agricultural and urban development, hydrologic changes associated with irrigated cultivation, overgrazing, and fire. Perhaps the best remaining glimpse into the original prairie can be found at Jepson Prairie, a 1,500 ac (600 ha) preserve in Solano County managed by the University of California, Davis.

Desert Grasslands

Grasslands occur at elevations of 3,500–5,700 ft (1,000–1,750 m) above the Sonoran and Chihuahuan deserts in the Basin and Range region of North America from southeastern Arizona to western Texas and south into northern Mexico. Not only is there more rain at these elevations than in the lower desert basins, but the cooler temperatures reduce evaporation so that precipitation becomes more effective and thus sufficient to support grasses and other plants that are excluded from the drier deserts below. These desert grasslands are composed primarily of perennial C_4 or warm-season grasses and leaf succulents. Grama grasses and tobosa grass dominate. Yuccas may dot the landscapes. On cooler north-facing slopes, woody plants such as junipers, live oaks, mesquite, and acacias may dot the landscape (see Figure 2.15). Where grazing by livestock has been heavy since their introduction in the mid-1700s, these woody plants have increased their presence throughout the desert grasslands.

A fairly high diversity of mammals is possible on the desert grasslands because of access to both desert habitat on lower slopes and forested habitat and permanently flowing springs on higher slopes. Among larger mammals are deer and javelina. Jackrabbit, desert cottontail, spotted ground squirrel, and kangaroo rats are among the smaller herbivores. Relatively few nesting birds are associated exclusively with these grasslands. Scaled, Gambel's, and Montezuma Quails are most notable.

The desert grasslands are utilized primarily for grazing cattle. Shrub invasion has been a major problem at least since the great cattle drives of the late 1800s.

The Eurasian Steppes

Natural grasslands, usually called steppes, stretch some 5,000 mi (8,000 km) across the Eurasian continent from eastern Europe into China. They form a continuous

Figure 2.15 Desert grasslands show the affect of aspect: north-facing slopes are dotted with oaks and other small trees, while drier south-facing slopes tend to lack shrubs and trees. *(Photo by author.)*

belt west of the Urals, but become fragmented in the large basins east of these mountains, which are an unofficial boundary between Europe and Asia (see Figure 2.16).

This huge vegetation belt is about 500 mi (800 km) wide in a north-south direction. Throughout much of the region elevation and relief are similar. (The steppes were a natural passageway between Asia and Europe along which invading armies and peoples readily moved in the past.) The parent material of soils is essentially uniform; much of it is loess.

Unlike the North American prairies, latitude plays a key role in differentiating various types of steppe and soils in Eurasia. The main factor that changes from north to south is total annual precipitation, it being higher in the north and lower toward the south. The northern steppe region is a zone of transition between the forest and the steppe known as forest-steppe, where meadow-steppe alternates with patches of forest. Parallel to forest-steppe and immediately south of it lies true steppe, a vegetation rich in perennial forbs and grasses. The next zone is a drier steppe of drought-tolerant bunchgrasses and perennial forbs. Toward the southern limits of the biome, dry bunchgrass steppes are dotted with dwarf shrubs and grade into true desert.

The climate becomes drier east of the Urals due to that area's deep interior position on the world's largest continent. Temperatures in eastern basins are more extreme than in European parts of the biome, also because of the effects of geographic location in the deep interior of the continent.

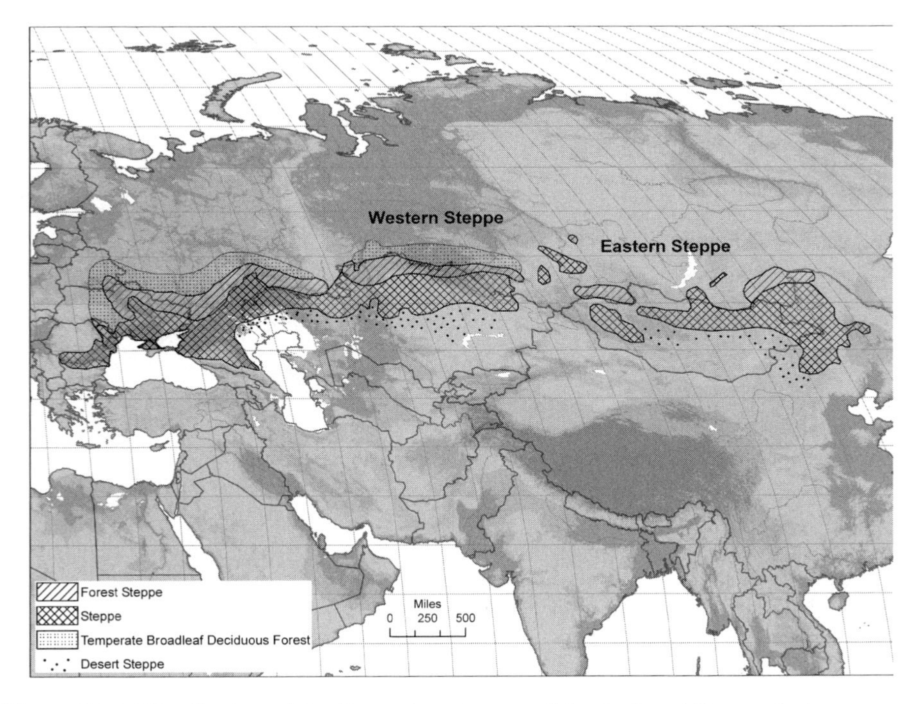

Figure 2.16 Distribution of the Eursasian steppes. *(Map by Bernd Kuennecke.)*

Throughout the steppes, the dominant grasses are perennial bunchgrasses. These are especially well adapted to withstand the cold temperatures and late summer drought that characterize this part of the biome, where the growing season is only about four months long. Feathergrasses of the genus *Stipa* are most common, and their long wispy seed heads lend a softness to the steppe landscape (see Figure 2.17). The feathergrasses are able to regulate water loss by rolling their leaves and closing their stomata when the weather is hot and dry. The actual feathergrasses present vary according to climate and geographic location.

Grasses become shorter as one moves from north to south in the biome. In the north, dominant grasses are typically 30–40 in (80–100 cm) tall, whereas in the dry grasslands of the southernmost parts of the biome they are only 6–8 in (15–20 cm) high. The degree to which grasses cover the ground also decreases as one goes from north to south. In northern meadow-steppe, grass cover is 70–90 percent. Plant cover in the driest areas is only 10–20 percent.

Particularly in the moister western steppes, perennial forbs are both abundant and diverse. Their flowers paint the steppes in ever-changing hues as species after species come into bloom during spring and early summer. As many as 11 color phases or "aspects" have been described in moister parts of the biome. Four or five aspects are more typical of the drier grasslands.

The short growing season also represents a short period of time in which the decay of dead plant material above and below ground can take place. The result is

Figure 2.17 Feathergrasses are conspicuous elements of the steppe near Kursk, Russia. *(Photo by author.)*

a deep accumulation of humus in the soil and the development of mollisols. The term "mollisol" is not used in Eurasia, where they have their own soil classification systems. Instead they are generally referred to as chernozems.

Soils generally reflect the climate pattern; and they, too, appear in distinct zones running from north to south (see Table 2.3). Northern chernozems underlie damp meadow-steppes in the forest steppe. South of this is a zone of thick chernozem under true steppe, a rich mixture of perennial forbs and grasses. South of that, normal chernozems have developed under a feathergrass steppe with abundant forbs. Finally, toward the southern limits of the steppes, are southern chernozems, which are produced in dry dwarf shrub-bunchgrass steppes.

The short time available for decay each year can lead to a buildup of litter on the surface if there is no other means available to remove dead plant matter. Too much litter prevents regeneration of grasses and opens space for weedy invaders such as wormwoods (*Artemisia* spp.) and knapweeds or star-thistles (*Centaurea* spp.). Originally wild ungulates grazed the steppes and prevented the accumulation of overabundant dead material. Occasional wild fires would have helped eliminate litter. Today, the larger herbivores have all been exterminated, and the steppes must be managed by mowing to conserve the cover of the native grasses and forbs.

As in temperate grasslands elsewhere, burrowing rodents that live in colonies are abundant. Members of the hare family (lagomorphs) are also present. These small mammals do not hibernate, but store forage to eat during the winter months. In the summer, they feed on green foliage, the succulent bases of grasses, rhizomes, bulbs, and tubers. Their populations tend to build up and then crash on four- to five-year cycles.

The Ural Mountains represent a boundary not only between climates but also between many plant and animal species' distribution areas. The more detailed descriptions below separate the Eurasian steppe biome into a western (or East European) section and an eastern (or Asian) section.

Western Eurasian Steppes

The north-south zonation of the steppes is apparent in the western reaches of the biome (see Table 2.3). In the forest-steppe, moist meadow-steppe dominated by the feathergrass *Stipa pennata* alternates with forest patches of oak (*Quercus robur*) on

Table 2.3 Latitudinal Zones of Steppe Vegetation and Soils, Western Eurasia

	Vegetation Type	Soil
North	Forest-steppe: meadow-steppe	Northern chernozem
↓	True or typical-steppe	Thick chernozem
	Dry *Stipa* bunchgrass-steppe with forbs	Normal chernozem
South	Desert shrub: bunchgrass-steppe	Southern chernozem

loess-based northern chernozems. Elsewhere, other feathergrasses and grasses from other genera are more important. Forbs bloom sequentially during the growing season, giving rise to the changing color aspects typical of these steppes. Grasses grow throughout the growing season.

As on the North American prairies, the native large mammals consisted of herds of ungulates. A small wild horse, the tarpan, and the ancestor of domestic cattle, the aurochs, were the major herbivores, but both are extinct in the wild today.

Burrowing rodents are abundant and diverse and responsible to a great degree for turning over the soil much as earthworms do. Their activity brings nutrients leached to depth back to the surface and helps maintain soil fertility. One of the more interesting rodents is the mole rat, which until recently was thought to belong to a unique family found only in Eurasia. These colonial animals spend their entire lives tunneling below ground, but the landscape is pimpled with the 6–12 in high (2–5 cm high) mounds of dirt that they have cleared from their passages.

The Many Aspects of the Flowering Steppes

The steppes turn color as different wildflowers bloom between early April and late June. Earliest spring sees a brown phase created by the blossoming of pasqueflowers and the bunch sedges. By late April, the steppes are yellow with sweet vernal (see Plate V). In mid-May grasses begin to grow. Forget-me-nots turn the steppes blue toward the end of May, although white snowdrop and anemone and yellow field fleaworts are also in flower. By June, when clover, Shasta daisies, and dropwort bloom, the white aspect is under way. The number and timing of aspects varies from year to year depending on weather conditions.

The squirrel family is represented by suslik, bobak, and tarbagan. Despite the strange-sounding names, these are close relatives of the ground squirrels and marmots of the North American prairies. Also inhabiting the steppes is a pika, a relative of hares and rabbits.

The western Eurasian steppes became the breadbasket of Ukraine, Russia, and other countries whose territory overlaps the biome. Plowing destroyed the native grass cover beginning in the 1940s. Today, the rich soils, though subject to periodic severe droughts, produce mostly wheat, sugar beets, and buckwheat.

Eastern Eurasian Steppes

The Asian steppes are under the influence of the changing pressure and wind systems of the Asian monsoon, and plants and animals are adapted to a more severe climate both in terms of less total precipitation and much colder and drier winters than on the European side of the mountains. In western Siberia, the plants are similar to those found on the East European steppe, except for the prominence of bulbs such as lilies. East of Lake Baikal, in Mongolia and China, the steppe occurs in isolated basins and valleys separated from each other by high mountain ranges. With little or no winter precipitation, spring wildflowers are absent from places such as Mongolia. Most plants bloom between June and August, when some moisture may arrive with the summer monsoon.

Chinese steppes have an east-west zonation similar to North American prairie and unlike the steppes of western Eurasia. Chinese scientists recognize three types of temperate grassland, the wettest in the east and the driest in the west of the steppe region. They refer to them as meadow-steppe, typical-steppe, and desert-steppe, respectively. Meadow-steppe is limited to elevations of 400–800 ft (120–250 m) in China's Northeast (formerly Manchuria), where it occurs on dark chestnut soils or chernozem. The dominant bunchgrass, *Stipa baicalensis*, reaches heights of 24 in (60 cm) when mature. Several sod-forming grasses are subdominants. A large number of perennial forbs grow in this part of the Chinese steppe. Tansy is most prevalent; daylilies and delphiniums are also common.

Typical-steppe occurs from the Northeast onto the inner Mongolian Plateau and south to the Loess Plateau. Elevations in this area range from 2,600–4,600 ft (800–1400 m). Annual precipitation is 12–16 in (300–400 mm). Grasses are shorter than in meadow-steppe and attain mature heights of 12–20 in (30–50 cm). Tufted feathergrasses dominate other tuft grasses such as cleistogenes, junegrasses, and bluegrasses. Perennial forbs are few, but dwarf shrubs such as wormwood are abundant.

Desert-steppe covers the southern Gobi and the northern and northwestern parts of the Loess Plateau. Plants are no taller that 8–12 in (20–30 cm) and widely spaced, so there is much bare ground. The drought-adapted tuftgrass *Stipa globica* is the dominant grass. Mongolian wormwood (*Artemisia xerophytica*), with a stem 2–3 in (5–8 cm) high, is the dominant woody plant. There are some bulbs related to onions (genus *Allium*) and some annual forbs—two growthforms that become more prevalent the drier the environment becomes.

The eastern Eurasian steppes are still home to the takhi or Przewalski's horse and the two-humped Bactrian camel, though the numbers of both are extremely low. The camels may or not may not be truly wild; it is likely that many, if not most, are feral. Takhi are partly the product of twentieth-century captive breeding and release programs.

In addition to these ungulates Asiatic wild asses or kulans and saiga antelope still roam in remote typical-steppe areas, though they are becoming increasingly rare. On desert-steppe, the Mongolian gazelle and goitered gazelle can be found. Burrowing rodents that live in colonies are common in the eastern steppes as they are in other temperate grasslands. The hairy-footed hamster is a resident of typical-steppe; its close relative the desert hamster is found in desert-steppe. Reptiles and amphibians are few in kind.

The eastern Eurasian steppes have suffered from millennia of human activity, especially the grazing of the flocks of nomadic herdsmen and, in moister areas, the cultivation of crops. The Mongolian steppes may be closest to original condition because burning and mowing are rare and grazing not overly heavy (see Figure 2.18).

Figure 2.18 Mongolian steppe at the base of the Altay Mountains. *(Photo courtesy of Dr. Sandra Olsen, Carnegie Museum of Natural History.)*

South American Temperate Grasslands

The South American Pampas

Flat to gently rolling lands on both sides of the Rio de la Plata of South America are another major area of temperate grassland (see Figure 2.19). These treeless grasslands generally are known as pampas. Two types are recognized: the pampas proper in Argentina west of the Rio de la Plata and Rio Uruguay and the Uruguayan pampas that extend southward from Brazil through Uruguay on the east of the two rivers. The Uruguayan pampas are also called—from a Brazilian perspective—the southern *campos*. To more readily distinguish between these two types of pampa, the term southern campos is used here to refer to the southern part of the biome.

Total annual precipitation ranges from 16 in (400 mm) in the southwestern pampa of Argentina to as much as 63 in (1,600 mm) in the northeast in the Brazilian state of Rio Grande do Sul. The northeastern areas, in particular, receive annual precipitation in amounts sufficient to support forests, and it is not entirely understood why the vegetation is dominated by grasses instead of trees. Temperatures are mild throughout the year compared with Northern Hemisphere prairie and steppe climatic regimes. Subfreezing temperatures may occur 125 days a year in the inland, western part of the biome, but only 20 days a year in the east, where

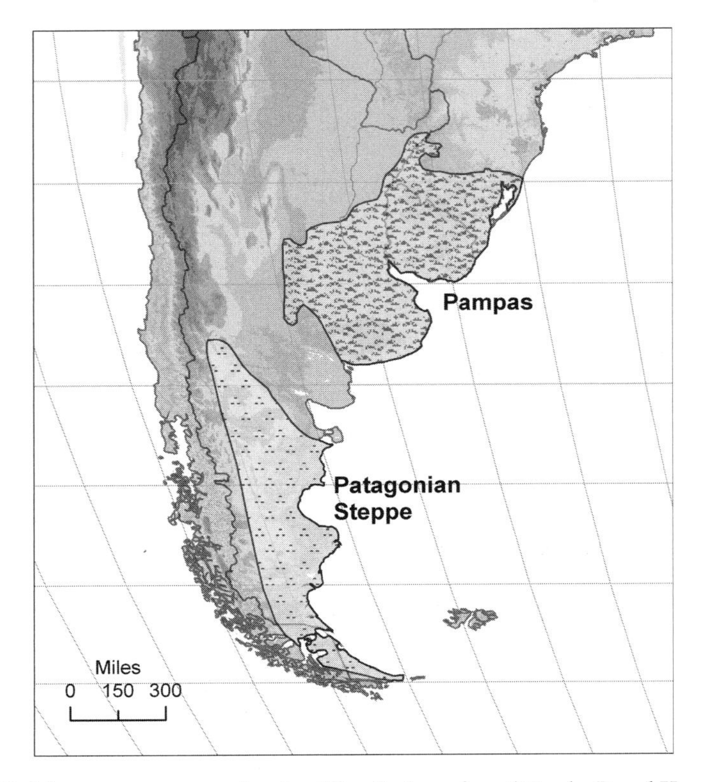

Figure 2.19 The temperate grasslands of South America. *(Map by Bernd Kuennecke.)*

the moderating influence of the sea is apparent. The length of the growing season varies from four to more than eight months.

The pampas lie between latitudes 23° and 40° S. The area receives moist air from a subtropical high pressure system in the South Atlantic. Its location in the Southern Hemisphere's zone of prevailing westerly winds brings dry winds off the Andes to the region south of about 34° S. As a result, the climate is humid in the east near the coast and becomes drier and drier to the west until, near the foot of the Andes, true desert is encountered.

Soils in the pampas proper are mollisols that have developed on loess or deep alluvium derived from loess. Moister eastern areas have chernozems, and the drier western pampas have chestnut soils. In the more humid southern campos, however, the pampa soils developed on a substrate of sand dunes and are alfisols, a soil order more usually associated with the Temperate Broadleaf Deciduous Forest Biome.

Grasses of the pampas are bunchgrasses and tussock grasses. Dominant grasses are tall to medium height. Almost all are warm-season C_4 grasses. Most are widespread, but dominance changes from place to place. Many of the perennial forbs and subshrubs are from the sunflower and pea families.

The pampas proper lies in the Argentine provinces of Formosa, La Pampa, Entre Rios, Corrientes, Santa Fe, and Buenos Aires, with the largest part in the Province of Buenos Aires. Several types of pampa distribute themselves according to precipitation totals. Between the Rio de la Plata and the Rio Paraná is a species-rich type known as rolling pampa. Plants here may become 20–40 in (50–100 cm) tall. The grasses are both sod-forming types and bunchgrasses. They grow to different heights, giving a layered structure to the vegetation. The tallest and dominant grass is the C_4 sod-forming silver beardgrass. Chilean needlegrass, a C_3 bunchgrass, grows to about 20 in (50 cm) and is part of the middle layer. A number of short tuftgrasses, about 4 in (5 cm) high and known locally as little darts or *flechillas,* make up the bottom layer. Sedges and small forbs also grow in the lowest layer. Grazing by domesticated animals has eliminated the tall grasses and reduced the vegetation to two layers.

East of the Rio de la Plata in the southern campos different grasses are found, but a distinct three-layered vegetation similar to the rolling pampa and the North American tall-grass prairie still is present. Tall grasses such as bahia grass and caninha grass are dominant. Fewer types of short grasses and more genera of legumes are represented than in the pampas proper. Under very wet conditions, tall plumes of Pampas grass wave 8 ft (2.5 m) above the ground.

In the dry southwestern part of the pampas proper and on some poor sandy soils in the north, a tussock grassland has developed. The tussock growthform is not found among Northern Hemisphere temperate grasslands and consists of a cluster of grass stems and leaves in which dead and fresh leaves occur in the same bunch. Individual tussocks may be more than 3 ft (1 m) tall. The presence of the dry, hard, dead blades gives these cool-season grasses a yellowed look throughout the year. The most common grasses were two feathergrasses: punagrass and serrated tussock grass. They had no value as forage for livestock, so most have been plowed under. The area has been replanted with more nutritious nonnative grasses.

Farther west, closer to the Andes, where precipitation declines to about 20 in (500 mm) a year, the pampas grade into a mesquite (*Prosopis* spp.) and acacia (*Acacia* spp.) woodland known as *monte,* and finally into a creosotebush (*Larrea divaricata*) desert where total rainfall reaches about 8 in (200 mm) a year.

South American pampas seem to have lacked the herds of large grazing animals characteristic of Northern Hemisphere grasslands. The absence may be related to the presumed young age of the vegetation type. Some scientists believe repeated burning by native people may have destroyed earlier forests and led to their replacement by grasses. According to this argument, there has not been enough time for a distinctive group of large mammals to evolve in adaptation to the pampas environment.

The only large mammalian plant eater was the pampas deer, now preserved in a few wildlife sanctuaries. A considerable variety of South American or neotropical rodents do inhabit the pampas, including the large plains viscacha that lives in colonies and digs intricate systems of burrows.

Smaller field mice, vesper mice, burrowing mice, and tuco-tucos represent other neotropical rodent genera, as do wild guinea pigs. The Patagonian hare or mara, also a rodent, is now rare. Other small mammals include several marsupials (opossum) and several armadillos—members of an essentially neotropical order, Xenarthra. The larger native carnivores, such as jaguar and puma, have been eliminated. Smaller carnivores such as the pampas fox, the ferret-like grisón, the pampas cat, and Geoffroy's cat are rare.

South America is known as the Bird Continent, so it is not surprising that a large number of birds inhabit the pampas. The flightless Rhea is the largest. Tinamous, doves, and parrots are also abundant. Among predatory birds are the Ornate Hawk-Eagle and the Crested Caracara.

Much of the pampas has been put under cultivation. This is especially true for the black soil areas of Buenos Aires Province, Argentina. Today, domesticated annual grasses such as maize, sorghum, wheat, and barley dominate. Alfalfa is produced for livestock; cattle raising—for beef and dairy products—remains the chief land use in the region. Areas not cultivated are rangeland. Grazing and the repeated burning of the pampas to stimulate the regrowth of grass shoots have altered the composition of the plant community to the extent that even the remnants are considered seminatural at best. The more palatable and taller grasses such as bahia grass have declined or disappeared, and drought-resistant woody plants, especially chanar or Chilean palo verde (*Geoffroea decorticans*), have invaded. In their degraded condition, the pampas are again similar to tall-grass prairie in mid-continental North America. The spread of exotic plants is, similarly, also a problem.

The Patagonian Steppe

In southern Argentina and southeastern Chile between the latitudes of 40° and 52° S lie Patagonia and Tierra del Fuego, cold semiarid regions in the rainshadow of the southern Andes Mountains (see Figure 2.19). Annual precipitation is less than 10 in (250 mm) a year, but cool temperatures keep evaporation rates so low that there is sufficient moisture to support grasses. Temperatures are cold year-round, but not severely so. The tapered cone shape of southern South America and winds blowing down the slopes of the Andes usually keep winter temperatures at or near freezing. Constant strong winds are more of a limiting factor for plant growth than low temperatures.

Climate data are scarce for this part of South America. A site in Chubut Province, Argentina, at 45° S, 70° W, and described as typical for the Patagonian steppe, receives about 6 in (150 mm) of precipitation annually, most during the Southern Hemisphere fall and winter months. The mean monthly temperature for the coldest month at this location, July, is 33° F (2° C), while temperatures for the warmest month, January, average only 40° F (14° C).

Soils are coarsely textured with a high gravel content. Calcium carbonate concentrates in a layer some 18–24 in (45–60 cm) below the surface.

Low tussocks of perennial feathergrasses and fescues known locally as different kinds of *coirón* are dominant (see Figure 2.20). The tussock grasses grow 12–24 in (30–70 cm) high. They often cover only about 30 percent of the ground. Interspersed among the tussocks are cushion plants such as *nuneo* and other globe-shaped woody shrubs. The cushion growthform, in which a many-stemmed shrub grows low to the ground as a dense mound, provides protection against drying winds for the renewal buds. Stubby-branched shrubs such as *mata negra* and dwarf shrubs such as *colapiche* are also characteristic.

The tussock grasses have shallow root systems and maintain green blades all year long. The shrubs have deep root systems and become dormant in winter. There tend to be two patterns to the way plants cover the landscape: individual tussocks may be scattered over largely bare ground, or shrubs may be enclosed in a close ring of grasses.

Animal life on the Patagonian steppe is quite diverse. Among herbivorous mammals, many of which are rare or threatened with extinction today, are guanaco (see Plate VI), mara (see Figure 2.21), and viscacha. Carnivores include foxes, Patagonian weasel, and puma.

Many different kinds of birds live on the steppe. Many are endemic. Darwin's or the Lesser Rhea inhabits these steppes (see Plate VII), as does the Patagonian Tinamou, Patagonian Mockingbird, and Patagonian Yellow-Finch.

Figure 2.20 Tussock grasses dominate the Patagonian steppe. *(Photo © Theo Allofs/ CORBIS.)*

Figure 2.21 Mara or Patagonian hare. *(Photo © Stephen Meese/Shutterstock.)*

The Patagonian steppe is sparsely inhabited by people, yet there has been serious degradation of the vegetation and consequent desertification. Overgrazing by sheep, the chief culprit, has exposed the soil to erosion. Perhaps the place where this type of vegetation is most often seen by outsiders is in Torres del Paine National Park, Chile.

Patagonian Hare

The so-called Patagonian hare is not a lagomorph like the jackrabbits and other true hares of the Northern Hemisphere, but one of South America's many caviomorph rodents. It is perhaps better to call it by the Araucanian name, mara.

Two species are known. The larger, *Dolichitos patagonum*, inhabits central and southern Argentina. The combined length of its head and body is about 28 in (690–750 mm) and it weighs 20–35 lbs (9–16 kg). A smaller or dwarf form (*D. salinicola*), which is only about 17 in (450 mm) in total length, occupies grasslands from southern Bolivia through Paraguay into northern Argentina.

The maras demonstrate some of the characteristics of lagomorphs and ungulates in northern prairies and steppes. They have long legs for running, though they also hop and stott. Diurnal creatures, they seek shelter underground at night either in burrows they dig themselves or in those abandoned by viscachas.

Darwin remarked in 1833 that in Patagonia "it is a common feature in the landscape to see two or three hopping quickly one after the other in a straight line across these wild plains."

African Temperate Grasslands

The South African Veld

The temperate grasslands or veld of South Africa occurs at high elevations in subtropical latitudes. The term veld actually means vegetation, though it has come to be used by North American and European scientists as a synonym for temperate grassland. In South Africa, people distinguish the temperate grassland from other vegetation types by calling it highveld. Highveld is centered on the high rolling plateaus and the eastern slopes of mountains in the east-central part of South Africa and western Lesotho at elevations of 4,000–6,000 ft (1300–2000 m) above sea level (see Figure 2.22 and Plate VIII).

Rainfall averages from 20–30 in (500–800 mm) a year. It is lowest in Eastern Cape and highest in Free State and occurs mostly during the summer months (September through April). Summers can be quite warm with average January temperatures near 70° F (21° C). Severe frosts are common in winter, when temperatures can plunge to −5° F (−15° C). Snow falls in the higher elevations.

Soils are influenced by the bedrock of shales and sandstones. They tend to be deep and red or yellow in color. In areas where the soils developed on volcanic

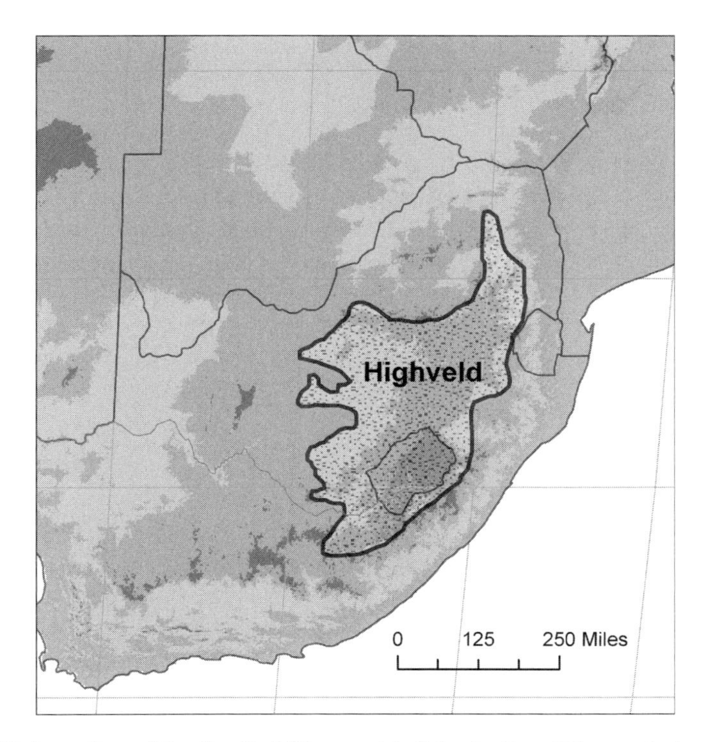

Figure 2.22 Location of the South African veld. *(Map by Bernd Kuennecke.)*

Figure 2.23 Redgrass, a common grass of the highveld. *(Photo by author.)*

material, however, they tend to be more like the dark, fertile mollisols of the Northern Hemisphere prairies and steppes.

More than 50 species of grass are reported from South Africa's highveld. At lower elevations, where winters are milder, a tussock grassland is dominated by redgrass (see Figure 2.23) almost to the exclusion of other grasses.

This is a highly nutritious warm-season (C_4) grass. Forbs are rare. Cool-season (C_3) grasses are more abundant in the more species-rich grasslands found at higher elevations. Bushveld turpentinegrass and small creeping foxtail are codominants with redgrass. Many more forbs also grow at higher elevations. South African grasslands have an extremely high biodiversity, second in the country only to the mediterranean fynbos of the Cape Floristic Province.

Unlike temperate grasslands in other parts of the world, the veld is home to a large number of ungulates. Many of the animals are the same ones found in Africa's Tropical Savanna Biome. Antelopes and zebras, baboons and vervet monkeys,

Sweet and Sour Grasses

A distinction is made in southern Africa between sweet grasses and sour grasses in the veld. The sweet grasses, such as redgrass, are those most palatable to livestock because they have low fiber content and maintain nutrients in their leaves during the winter when they are dormant. Sweetveld is found in areas with richer soils developed on shales and volcanic rocks and annual precipitation usually less than 25 in (625 mm). Sourveld develops in areas of higher elevation where there are acidic soils and more rainfall. The grasses of the sourveld withdraw nutrients from their leaves in winter and have a higher amount of fiber than sweet grasses, so they are of less value as forage, especially during the nongrowing season.

Figure 2.24 Cape mountain zebra in highveld, Mountain Zebra National Park, South Africa. *(Photo © Chris Kruger/Shutterstock.)*

springhares and striped mice, porcupines, jackals, caracals, and mongooses all live in these grasslands. Only a few animals are restricted to these high elevation grasslands. The black wildebeest, smaller cousin of the more widespread blue wildebeest, grazes among more common antelope species, such as red hartebeest, blesbok, springbok, and steenbok. The Cape mountain zebra (see Figure 2.24), a distinct subspecies of mountain zebra, feeds on coarse grasses at elevations up to 6,000 ft (2,000 m), where its populations are dangerously small. Recognized by its obvious dewlap, this zebra is one of the most endangered mammals in the world. The much more common Burchell's or plains zebra also inhabits highveld.

Large cats have been largely extirpated, but cheetahs have been reintroduced in national parks. Other carnivores are relatively small and include caracals, aardvarks (a hyena), black-tailed jackals, and yellow- and white-tailed mongooses. Birdlife is diverse and includes ground-dwelling birds such as francolins and brightly colored songsters such as Orange-breasted Longclaws. A lucky bird watcher might spy the very rare and endemic Bearded Vulture soaring overhead.

Ancient human impacts are as yet unknown, but the oldest Khoisan rock paintings in the region may be 25,000 years old. Pastoralist peoples arrived much more recently with their beautifully spotted Nguni cattle. Still more recently, European settlers found a cool area free of the tropical diseases that affect livestock and turned the area into rangelands for their own breeds of cattle and sheep. Soils and climate were well suited to Northern Hemisphere crops such as wheat and maize, so the veld was sectioned off into farms and plowed under. Only a few fragments of natural grassland remain and many of these are threatened by overgrazing. Mountain Zebra National Park and Golden Gate Highlands National Park are two areas where efforts to preserve, restore, and manage highveld communities are ongoing.

Further Readings

Books

Hoogland, John L. 2006. *Conservation of the Black-tailed Prairie Dog: Saving North America's Western Grasslands.* Washington, DC: Island Press.

Moul, Francis. 2006. *The National Grasslands.* Lincoln: University of Nebraska Press.

Internet Sources

National Geographic and World Wildlife Fund. 2001. "Nearctic Temperate Grasslands, Savannas and Shrublands." Wild World, Terrestrial Ecoregions of the World. http://www.nationalgeographic.com/wildworld/profiles/terrestrial_na.html#tempgrass.

Northern Prairie Wildlife Research Center. n.d. "Regional Trends of Biological Resources—Grasslands." http://www.npwrc.usgs.gov/resource/habitat/grlands.

Palouse Prairie Foundation. n.d. http://www.palouseprairie.org.

South African National Parks. n.d. "Mountain Zebra National Park." http//www.sanparks.org/parks/mountain_zebra.

Wind Cave National Park. n.d. "Grasses of the Mixed Grass Prairie." http://www.nps.gov/archive/wica/Grasses_of_the_Mixed_Grass_Prairie.htm.

Climate Data

WorldClimate. 1996–2008. Climate data. http://www.worldclimate.com.

Representative Stations:

North America

California prairie (northern section): Sacramento, CA
California prairie (southern section): Fresno, CA
Desert grasslands: Santa Rita Experimental Range, AZ
Mixed-grass prairie (northern section): Regina, Saskatchewan, Canada
Mixed-grass prairie (southern region): Abilene, TX
Palouse prairie (eastern section): Moscow, ID
Palouse prairie (western section): Pullman, WA
Short-grass prairie: Amarillo, TX
Tall-grass prairie: Topeka, KS

Eurasia

Eastern dry steppes: Semey, Kazakhstan
Forest steppe: Kursk, Russia
True steppe: Rostov-na-Donu, Russia

South America

Pampas: Buenos Aires, Argentina

Africa

Veld: Jan Smuts, South Africa and Middleburg, South Africa

Appendix

Selected Plants and Animals of the Temperate Grassland Biome

North American Temperate Grasslands

Some Characteristic Plants of the Tall-Grass Prairie

Grasses	
Tall grasses (about 6 ft at maturity)	
Big bluestem	*Andropogon gerardi*
Sand bluestem	*Andropogon hallii*
Switchgrass	*Panicum virgatum*
Mid-size grasses (about 4 ft at maturity)	
Little bluestem	*Schizachrium scoparium*[a]
Indiangrass	*Sorghastrum nutans*
Prairie sandreed	*Calamavilfa longifolia*
Short grasses (3 ft or less at maturity)	
Sideoats grama	*Bouteloua curtipendula*[a]
Forbs	
Sunflower family	
Yarrow	*Achillea* spp.
Purple coneflower	*Echinacea angustifolia*
Goldenrod	*Solidago* spp.
Sunflowers	*Helianthus* spp.
Pea family	
Wild alfalfa	*Psoralea tenuifolia*
Prairie clover	*Petalostemon spp.*
Slender lespedeza or bush clover	*Lespedeza virginica*

Note: [a]Bunchgrasses. The other grasses listed are all sod-forming types.

Some Characteristic Animals of the Tall-Grass Prairie

Large mammals	
Herbivores	
Bison	*Bison bison*[a]
Pronghorn	*Antilocapra americana*[a]
Elk	*Cervus canadensis*[a]
Mule deer	*Odocoileus hemionus*[a]
Carnivore	
Wolf	*Canis lupus*[a]
Small mammals	
Herbivores	
Eastern cottontail	*Sylvilagus floridanus*
White-tailed jackrabbit	*Lepus townsendii*
Thirteen-lined ground squirrel	*Spermophilus tridecemlineatus*
Meadow vole	*Microtus ochrogaster*
Plains harvest mouse	*Reithrodontomys montanus*
Grasshopper mouse	*Onychomys leucogaster*
Pocket gopher	*Geomys bursarius*
Carnivores	
Coyote	*Canis latrans*
Red fox	*Vulpes fulva*
American badger	*Taxidea taxus*
Least weasel	*Mustela frenata*
Striped skunk	*Mephitis mephitis*
Birds	
Upland Sandpiper	*Bartamia longicauda*
Burrowing Owl	*Speotypto acunicularia*
Greater Prairie Chicken	*Tympanuachus cupido*
Horned Lark	*Eremophila alpestris*
Grasshopper Sparrow	*Ammodramus savannarum*
Dickcissel	*Spiza americana*
Eastern Meadowlark	*Sturnella major*
Western Meadowlark	*Sturnella neglecta*

Note: [a]No longer present.

Some Characteristic Plants of the Mixed-Grass Prairie

Grasses	
Medium-height grasses (C_3)	
Western wheatgrass	*Andropogon smithii*
Sand bluestem	*Andropogon hallii*
Dropseed	*Sporobolus cryptandrus*

(*Continued*)

Junegrass	*Koeleria cristata*
Needle-and-thread	*Stipa comata*
Short grasses (C_4)	
Blue grama	*Bouteloua gracilis*
Buffalograss	*Buchloë dactykoides*
Forbs	
Western yarrow	*Achillea lanulosa*
Heath aster	*Aster ericoides*
Plains beebalm	*Monarda pectinata*
Woody plants	
Fringed sagebrush	*Artemisia frigida*
Silver sagebrush	*Artemisia cana*
Wild rose	*Rosa* spp.
Trembling aspen	*Populus tremuloides*[a]
Succulents	
Bitter prickly pear cactus	*Opuntia fragiles*
Plains prickly pear cactus	*Opuntia polycanta*

Note: [a]Recent invader since control of prairie fires.

Some Characteristic Animals of the Mixed-Grass Prairie

Large mammals	
Herbivores	
Bison	*Bison bison*[a]
Pronghorn	*Antilocapra americana*[a]
Elk	*Cervus canadensis*[a]
Mule deer	*Odocoileus hemionus*[a]
Carnivore	
Wolf	*Canis lupus*[a]
Small mammals	
Black-tailed prairie dog	*Cynomys ludovicianus*
Northern pocket mouse	*Thomomys talpoides*
Meadow jumping mouse	*Zapus hudsonius*
Birds	
Ferruginous Hawk	*Buteo regalis*
Golden Eagle	*Aquila chrysaetos*
Sharp-tailed Grouse	*Pedioectes phasianelles*
Baird's Sparrow	*Ammodramus bairdii*
Sprague's Pipit	*Anthus spragueii*
Chestnut-collared Longspur	*Calcarus ornatus*
McCown's Longspur	*Rhynchophanes mccownii*

Reptiles	
Prairie rattlesnake	*Crotalus viridus*
Plains hognose snake	*Opheodrys vernolis*
Bull snake	*Pituophis melanleuces*
Plains garter snake	*Thamnophis radix*
Short-horned lizard	*Phyrnosoma douglassi*
Amphibians	
Great Plains toad	*Bufo cognatus*
Leopard frog	*Rana pipiens*

Note: [a]Exterminated.

Some Characteristic Plants of the Short-Grass Prairie

Grasses	
Short grasses	
Blue grama	*Bouteloua gracilis*
Buffalograss	*Buchloë dactyloides*
Mid-size grasses	
Western wheatgrass	*Andropogon smithii*
Needle-and-thread grass	*Stipa comata*
Red three-awn	*Aristida longiseta*
Woody plants	
Fringed sagebrush (or prairie sagewort or wormwood)	*Artemisia frigida*
Silver sagebrush	*Artemisia cana*
Wild rose	*Rosa* spp.
Succulents	
Bitter prickly pear cactus	*Opuntia fragiles*
Plains prickly pear cactus	*Opuntia polycanta*

Some Characteristic Animals of the Short-Grass Prairie

Large mammals	
Herbivores	
Bison	*Bison bison*
Pronghorn	*Antilocapra americana*
Small mammals	
Herbivores	
Black-tailed jackrabbit	*Lepus californica*
White-tailed jackrabbit	*Lepus townsendii*
Desert cottontail	*Sylvilagus audubonii*
Thirteen-lined ground squirrel	*Spermophilus tridecemlineatus*

(*Continued*)

White-tailed prairie dog	*Cynomys leucurus*
Plains pocket gopher	*Geomys bursarius*
Northern pocket gopher	*Thomomys talpoides*
Carnivores	
Badger	*Taxidea taxus*
Black-footed ferret	*Mustela nigripes*[a]
Coyote	*Canis latrans*
Birds	
Mountain Plover	*Charadrius montana*
Burrowing Owl	*Speotypto acunicularia*
Greater Prairie Chicken	*Tympanuachus cupido*
Horned Lark	*Eremophila alpestris*
Brewer's Sparrow	*Spizella breweri*
Western Meadowlark	*Sturnella neglecta*
Lark Bunting	*Calamospiza melanocorys*
Chestnut-collared Longspur	*Calcarus ornatus*
McCown's Longspur	*Rhynchophanes mccownii*
Reptiles	
Prairie rattlesnake	*Crotalus viridus*
Gopher snake	*Pituophis catenifer*
Western hognose snake	*Heterodon nasicus*
Skinks	*Eumeces* spp.
Short-horned lizard	*Phrynosoma douglassi*
Box turtle	*Terrapene ornate*
Amphibians	
Great Plains toad	*Bufo cognatus*
Woodhouse's toad	*Bufo woodhousei*
Leopard frog	*Rana pipiens*

Note: [a]Endangered.

Some Characteristic Plants of the Palouse Prairie

Grasses	
Bluebunch wheatgrass	*Agropyron spicatum* (=*Pseudoroegneria spicata*)
Idaho fescue	*Festuca idahoensis*
Sandberg bluegrass	*Poa secunda*
Prairie junegrass	*Koeleria macrantha*
Bottlebrush squirreltail	*Elymus elymoides*
Perennial forbs	
Western yarrow	*Achillea lanulosa*
Yellow Indian paintbrush	*Castilleja lutescens*
Old man's whiskers	*Geum trifoleum*

Silky lupine	*Lupinus sericeus*
Brodiaea	*Brodiaea douglasii*
Lambstongue ragwort	*Senecio integerrimus*
Camas (wild hyacinth)	*Camassia quamas*
Annual forbs	
Willow herb	*Epilobium paniculatum*
Indian lettuce	*Montia linearis*
Shining chickweed	*Stellaria nitens*
Dwarf shrubs	
Arrowleaf balsamroot	*Balsamorhiza sagittata*
Common snowberry	*Symphoricarpus albus*
Wild roses	*Rosa nutkana, Rosa woodsii*

Some Characteristic Animals of the Palouse Prairie

Large mammals	
Herbivores	
Bison	*Bison bison*
Pronghorn	*Antilocapra americana*
Mule deer	*Odocoileus hemionus*
Whitetail deer	*Odocoileus virginianus*
Elk	*Cervus canadensis*
Small mammals	
Herbivores	
White-tailed jackrabbit	*Lepus townsendii*
Mountain cottontail	*Sylvilagus nuttalii*
Yellow-bellied marmot	*Marmota flaviventris*
Columbian ground squirrel	*Spermophilus columbianus*
Carnivores	
Coyote	*Canis latrans*
Badger	*Taxidea taxus*
Birds	
Brewer's Sparrow	*Spizella breweri*
Sharp-tailed Grouse	*Tympanuchus phasioanellus*
Reptiles	
Prairie rattlesnake	*Crotalus viridis*
Garter snake	*Thamnophis ordinalis*
Bull snake	*Pituophis catenifer*

Some Characteristic Plants of the California Prairie

Pre-1800	
Perennial grasses	
Purple needlegrass	*Stipa pulchra*
Malpais bluegrass	*Poa scabrella*
Three awn	*Aristida* spp.
Wild ryes	*Elymus glaucus, Leymus triticoides*[a]
Junegrasses	*Koeleria cristata*
Annual grasses	
Small fescue	*Vulpia microstachys*
Six-weeks fescue	*Vulpia octoflora*
Perennial forbs	
California poppy	*Eschsholtzia californica*
Purple owlsclover	*Orthocarpus purpurascens*
Post-1800	
Annual grasses	
Wild oats	*Avena fatua, Avena barbata*[b]
Soft brome	*Bromus hordeaceus*[b]
Annual forb	
Storksbill (filaree)	*Erodium botrys*[b]

Notes: [a]The only native sod-forming grasses; [b]Introduced.

Some Characteristic Animals of the California Prairies

Large mammals	
Herbivores	
Pronghorn	*Antilocapra americana*
Mule deer	*Odocoileus hemionus*
Tule elk	*Cervus canadensis nannodes*
Carnivores	
Wolf	*Canis lupus*
Grizzly bear	*Ursus horribilis*
Mountain lion	*Felis concolor*
Small mammals	
Herbivores	
California ground squirrel	*Citellus beecheyi*
Fresno kangaroo rat	*Dipodomys nitratoides exilis*
Tipton kangaroo rat	*Dipodomys nitratoides nitratoides*
San Joaquin antelope squirrel	*Ammospermophilus nelsoni*
San Joaquin pocket mouse	*Perognathus inornatus*
Giant kangaroo rat	*Dipodomys ingens*

Carnivores	
Bobcat	*Felis rufus*
Coyote	*Canis latrans*
San Joaquin Valley kit fox	*Vulpes macrotis mutica*
Reptiles	
San Joaquin whipsnake	*Masticophis flagellum ruddocki*
Blunt-nosed leopard lizard	*Gambelia sila*
Gilbert's skink	*Eumeces gilberti*

Some Characteristic Plants of the Desert Grasslands

Grasses	
Grama grasses	*Bouteloua* spp.
Toboso grass	*Hilaria mutica*
Plains lovegrass	*Eragrostis intermedia*
Mulhy	*Muhlenbergia* spp.
Leaf-succulent shrubs	
Yuccas	*Yucca* spp.
Sotol	*Dasylerion wheeleri*
Beargrass	*Nolina microcarpa*
Agaves	*Agave* spp.
Woody plants[a]	
Junipers	*Juniperus deppeana and Juniperus monosperma*
Emory oak	*Quercus emoryi*
Creosotebush	*Larrea tridentata*
Mesquites	*Prosopis* spp.
Ocotillo	*Fouqueria spendens*
Acacias	*Acacia* spp.

Note: [a]Invaders since the nineteenth century.

Some Characteristic Animals of the Desert Grasslands

Large mammals	
White-tail deer	*Odocoileus virginianus*
Mule deer	*Odocoileus hemionus*
Javelina or peccary	*Tayasssu tajuca*
Small mammals	
Black-tailed jackrabbit	*Lepus californicus*
Antelope jackrabbit	*Lepus alleni*
Spotted ground squirrel	*Spermophilus spilosoma*
Kangaroo rats	*Dipodomys ordii* and *Dipodomys spectabilis*

(*Continued*)

Coyote	*Canis latrans*
Badger	*Taxidea taxus*
Bobcat	*Felis rufus*
Gray fox	*Urocyon cineargenteus*
Kit fox	*Vulpes macrotis*
Striped skunk	*Mephitis mephitis*
Birds	
Scaled Quail	*Callipepla squamata*
Gambel's Quail	*Callipepla gambelli*
Montezuma Quail	*Crytonyx montezumae*

Eurasian Temperate Grasslands

Characteristic Plants of Western Eurasian Steppes

Grasses	
Feathergrasses	*Stipa lessingiana*, *Stipa pulcherrima*, and *Stipa zaleski*
Wheatgrasses	*Agropyron* spp.
Fescues	*Festuca* spp.
Oat grasses	*Helichtotrichon* spp.
Junegrasses	*Koeleria* spp.
Sedges	
Sedges	*Carex humilis* and others
Forbs	
Pasqueflower	*Pulsatilla patens*
Sweet vernal	*Adonis vernalis*
Forget-me-nots	*Myosotis sylvatica*
Snowdrop anemone	*Anemone sylvestris*
Field fleawort or groundsel	*Senecio campestris*
Clover	*Trifolium montanum*
Shasta daisy	*Chrysanthemum leucanthemum*
Dropwort	*Filipendula hexapetala*

Characteristic Animals of the Western Eurasian Steppes

Large mammals	
Tarpan	*Equus gmelini*[a]
Aurochs	*Bos taurus*[a]
Small mammals	
Suslik	*Spermophilus pygmaeus*
Bobak or steppe marmot	*Marmota bobak*

Tarbagan	*Marmota siberica*
Daurian pika	*Ochotonoa daurica*
Steppe lemming	*Lagurus lagurus*
Brandt's vole	*Lasiopodomys brandtii*
Chinese vole	*Lasiopopdomys mandarinus*
Common vole	*Microtus arvalis*
Narrow-skulled vole	*Microtus gregalis*
Social vole	*Microtus socialis*
Mole vole	*Ellobius talpinus*
Mole rat	*Spalax microphthalmus*

Note: [a]Extinct. The tarpan, a forest-dwelling species of the forest-steppe, has been recreated by back breeding, and a few zoos display them so that one can see what the original was like.

Plants of Eastern Eurasian True Steppes

Grasses	
Feathergrasses	*Stipa* spp.
Cleistogenes	*Cleistogenes squarrosa*, *Cleistogenes polyphylla*[a]
Junegrass	*Koerelia gracilis*
Bluegrass	*Poa sphondylodes*
Subshrub	
Fringed sagebrush or wormwood	*Artemisia frigida*

Note: [a]Genus endemic to Inner Mongolia.

Characteristic Animals of the Eastern Eurasian Steppes

Large mammals	
Kulan	*Equus hemionus*
Tahki	*Equus przewalskii*
Bactrian camel	*Camelus bactrianus*
Saiga	*Saiga tatarica*
Mongolian gazelle	*Procapra guffurosa*
Small mammals	
Suslik	*Spermophilus pygmaeus*
Bobak or steppe marmot	*Marmota bobak*
Hairy-footed hamster	*Phodopus sungorus*
Desert hamster	*Phodopus roborouskii*
Brandt's vole	*Lasiopodomys brandtii*
Narrow-skulled vole	*Microtus gregalis*
Zokor	*Myospalax aspalax*

Birds	
Horned Lark	*Eremophila alpestris*
Mongolian Lark	*Melanocorypha mongolica*
Skylark	*Alauda arvensis*
Isabelline Wheatear	*Oenanthe isabellina*
Common Wheatear	*Oenanthe oenanthe*
Reptiles	
Mongolian racerunner	*Eremias argus*
Steppe toad-headed agama	*Phrynocephalus frontalis*
Steppe ratsnake	*Elaphe dione*
Amphibians	
Siberian sand toad	*Bufo raddei*
Common frog	*Rana temporaria*

South American Temperate Grasslands

Some Characteristic Plants of the Pampas

The rolling pampa, Argentina	
Grasses	
Silver beardgrass	*Botriochlora laguroides* = *Andropogon laguroides*
Chilean needlegrass	*Stipa neesiana*
Uruguayan rice grass	*Piptochaetium montevidense*
"Little dart" or *flechilla*	*Aristida murina*
Plumerillo	*Stipa papposa*
Forbs	
Joe-pye-weed	*Eupatorium buniifolium*
Cat's ear or false dandelion	*Hypochoeris* spp.
Subshrub	
Carqueja	*Baccharis* spp.
The southern campos, Uruguay	
Grasses	
Caninha grass	*Andropogon lateralis*
Bahia grass (*pasto horqueta*)	*Paspalum notatum*
Pasto miel	*Paspalum dilatatum*
Bush beargrass	*Schizachyrium condensatum*
Carpet grass	*Axonopus compressus*
Brome (*cebadillo criolla*)	*Bromus catharaticus*
Pampas grass	*Cortaderia selloano*

Forbs	
Eryngium	*Eryngium* spp.
Goldenrod	*Solidago* spp.
Cudweed	*Gnaphalium* spp.
Kaimi clover	*Desmodium canum*
Tussock grasslands	
Grasses	
Punagrass	*Stipa brachychaeta*
Serrated tussockgrass	*Stipa trichotoma*

Some Characteristic Animals of the Pampas

Large mammal	
Pampas deer	*Ozotoceros bezoarticus*[a]
Small mammals	
Herbivores	
Opossum	*Didelphis azarae*
Opossum	*Lutreolina crassicaudata*
Armadillo	*Chaetophractus villosus*
Armadillo	*Chlamyphorus truncatus*
Armadillo	*Dasypus hybridus*
Plains viscacha	*Lagostomus maximus*
Patagonian hare or mara	*Dolichotis patagonum*
South American field mice	*Akodon* spp.
South American field mice	*Bolomys* spp.
Vesper mice	*Calomys* spp.
Rice rats	*Orysomys* spp.
Burrowing mice	*Oxymycterus* spp.
Water rats	*Scapteromys* spp.
Guinea pig	*Cavia* spp.
Cui	*Galea* spp.
Tuco-tucos	*Ctennomys* spp.
Carnivores	
Jaguar	*Panthera onca*[b]
Puma	*Felis concolor*[b]
Fox	*Psuedoalopex gymnocercus*
Grisón	*Galictis cuya*
Pampas cat	*Felis colcolo*
Geoffroy's cat	*Felis geoffroyi*
Birds	
Rhea	*Rhea americana*
Campo Flicker	*Colaptes campestris*

(*Continued*)

Ornate Hawk-Eagle	*Spizaetus ornatus*
Crested Caracara	*Polyborus pancus*

Notes: [a]Rare; confined to preserves; [b]No longer present.

Some Characteristic Plants of the Patagonian Steppe

Grasses	
Coirón amargo	*Stipa speciosa, Stipa humilis*
Stiff-leaved coirón	*Festuca gracillima*
Pasto hilo	*Poa lanuginosa*
Woody shrubs and cushion plants	
Neneo	*Mulinum spinosum*
Mata negra	*Junellia tridens*
Colapiche	*Nassauvia glomerulosa*
Calafate	*Berberis heterophylla*
Mamuel choique	*Adesmis ampestris*

Some Characteristic Animals of the Patagonian Steppe

Mammals	
Herbivores	
Patagonian opossum	*Lestodelphis halli*
Mara	*Dolichotis patagonum*
Southern viscacha	*Lagidium viscacia*
Wolffsohn's mountain viscacha	*Lagidium wolffsohni*
Guanaco	*Lama guanicoe*
Carnivores	
Puma	*Felis concolor*
South American gray fox	*Dusicyon griseus*
Humboldt's hog-nosed skunk	*Conepatus humboldti*
Patagonian weasel	*Lyncodon patagonicus*
Birds	
Grey Eagle-Hawk	*Geranoaetus melanoleucus*
Darwin's Rhea or Lesser Rhea	*Pterocnemia pennata*
Patagonian Tinamou	*Tinamotis ingoufi*
Patagonian Mockingbird	*Mimus patagonicus*
Patagonian Yellow-Finch	*Sicalis lebruni*
Reptiles	
Lizards	*Liolaemus fitzingeri, L. kingi*
Patagonian gecko	*Homonata darwinii*
Darwin's iguana	*Diplolaemus darwinii*

South African Temperate Grasslands

Some Characteristic Plants of the South African Veld

Grasses	
Redgrass	*Themeda triandra*
Bushveld turpentine grass	*Cymbopogon plurinodis*
Threadleaf bluestem	*Diheteropogon filifolius*
Tanglehead grass	*Heteropogon contortus*
Ngongoni bristlegrass	*Aristida junciformis*
Velvet signal grass	*Brachiaria serrata*
Pangola grass	*Digitaria eriantha*
Small creeping foxtail	*Setaria flabellata*
Junegrass	*Koeleria cristata*
Small oats grass	*Helictrichon turgidulum*
Tambookie grass	*Miscanthidium erectum*
Thatching grass	*Hyparrhenia hirta*
Forbs	
Veld everlasting	*Helichrysum rugulosum*
Groundsel	*Senecio erubescens*
Horseweed	*Conyza podocephala*
Daisies	*Berkheya onopordifolia, Berkheya pinnatifida*

Some Characteristic Animals of the South African Veld

Mammals	
Herbivores	
Cape mountain zebra	*Equus zebra zebra*[a]
Plains zebra (Burchell's zebra)	*Equus bruchelli*
Black wildebeest	*Connochaetes gnou*
Eland	*Taurotragus oryx*
Red hartebeest	*Alcelaphus buselaphus*
Blesbok	*Damaliscus dorcas*
Springbok	*Antidorcas marsupialis*
Steenbok	*Raphicerus campestris*
Grey duiker	*Sylvicapra grimmia*
Scrub hare	*Lepus saxatilis*
Cape hare	*Lepus capensis*
African porcupine	*Hystrix africaeaustralis*
Highveld gerbil	*Tatera bransti*
Striped mouse	*Rhabdomys pumilo*
Carnivores	
Aardwolf	*Proteles cristatus*

(*Continued*)

Caracal or African lynx	*Felis caracal*
Black-backed jackal	*Canis mesomelas*
Yellow mongoose	*Cynictis penicellata*
White tailed mongoose	*Ichneumia albicauda*
Omnivores	
Chacma baboon	*Papio cynocephalus ursinus*
Vervet monkey	*Chlorocebus aethiops*
Birds	
Black Stork	*Ciconia nigra*
Bearded Vulture	*Gypaetus barbatus*
Cape Vulture	*Gyps coprotheres*
Grey-winged Franklin	*Scleroptila africanus*
Orange-throated Longclaw	*Macronyx capensis*
Stonechat	*Saxicola torquatus*
Drakensberg Siskin	*Pseudochloroptila symonsi*

Note: [a]Endemic to this biome and today one of the most endangered mammals on Earth.

Conversion from VATS to a thoracotomy is not a sign of failure.

Complications after video-assisted thoracic surgery lobectomy

Published series show that VATS lobectomy has gained international acceptance [1–4]; however, less than 10% of lobectomies are currently performed with VATS because most thoracic surgeons are still not comfortable with the technique. The author and colleagues' experience with 1,100 VATS lobectomies, pneumonectomies, and segmentectomies over a 12-year period showed a mean length of hospital stay at 4.78 days and median length of hospital stay at 3 days. The mortality rate was 0.8%, and no complications occurred in 84.7% of patients [1]. These results are better than published results for lobectomy via thoracotomy.

There are now many single institutional, observational series that report VATS lobectomy to be a safe and reasonable procedure. In series involving 106 to 1,100 patients, the mortality rates varied from 0% to 2.6% [1–4]. Table 1 shows typical complications after a VATS lobectomy.

Key points: complications and acceptance

VATS lobectomy is slowly becoming a well-accepted procedure internationally.

Video-assisted thoracic surgery versus thoracotomy for lobectomy

The evidence is mounting that a VATS lobectomy may have advantages over a lobectomy by thoracotomy. Opponents believe that a VATS lobectomy is unsafe, an incomplete cancer operation, and offers no advantage over a thoracotomy for lobectomy. Proponents believe that VATS lobectomy is a safe and effective treatment for lung cancer. The medical literature supports the latter position, and the number of surgeons who hold the negative belief is decreasing.

As the literature shows benefit to the procedure and as the public demands minimally invasive operations, the momentum for VATS lobectomy is clearly growing. There are a few small, randomized studies. Two small randomized trials in the 1990s showed a small advantage for the VATS approach [5,6]; more recently, in a small randomized, single institutional series, Hoksch and colleagues [7] showed a complication rate of 50% after thoracotomy versus 18% after VATS.

Although there is no large, multi-institutional randomized, prospective study to compare VATS and thoracotomy approaches for lobectomy, observational series with similar cohorts suggest that there is short-term benefit to a VATS approach, without compromising the long-term survival [1–6]. Koizumi and colleagues [8] reported mortality rates for patients aged 80 to 91 were 20% for thoracotomy and 5.9% for VATS. A recent meta-analysis and systematic review of controlled trials showed considerable benefit during the hospital stay and recovery phase for VATS over thoracotomy [9].

Table 1
Typical complications after video-assisted thoracic surgery anatomic resections

Major complications		Minor complications	
Readmission	1%–2%	Atrial fibrillation	3%–12%
Pneumonia	2%	Air leak	5%
Myocardial infarction	1%	Transfusion	<5
Empyema	<1%	Serous drainage	<2%
Broncho pleural fistula	<1%	SQ emphysema	1%
Stroke	<1%	Gastrointestinal	<1%

Some patients experience more than one complication.
Abbreviations: Air leak, air leak lasting more than 7 days; Gastrointestinal complications, Ogilvie's syndrome, gastrointestinal bleed; Serous drainage, serous drainage requiring chest tube drainage for more than 7 days; SQ emphysema, subcutaneous air requiring reinsertion of chest tube or subcutaneous catheters.

Hospital stay

During the hospital stay, there appear to be many benefits for the VATS approach when compared with a thoracotomy. The overall hospital length of stay was reduced by 2.6 days for the VATS approach [9]. This difference was particularly pronounced for elderly patients (length of stay 5.3 ± 3.7 versus 12.2 ± 11.1 days, $P = .02$ and duration of chest tubes 4.0 ± 2.8 versus 8.3 ± 8.9 days, $P = .06$). The charges for laboratory examinations, anesthesia, disposable equipment, and hospitalization were significantly higher in patients who underwent open thoracotomy, compared with the patients who underwent VATS [10].

Both randomized and controlled studies showed a 48% reduction in overall risk of complications [9]. Pulmonary complications, including respiratory dysfunction, pneumonia, atelectasis, empyema, and prolonged air leak, were reduced [9]. Although the incidence of atrial fibrillation in the author and colleagues' series of 1,100 patients was less than 3% [1], the meta-analysis found no difference in the incidence of cardiac complications, including atrial fibrillation [9]. The incidence of blood loss greater than 500 mL in a case was no different, but the VATS patients did have a significant reduction in blood loss [9]. The conversion rate from VATS to thoracotomy was 6% [9]. In most cases, conversion was necessary for oncologic (eg, need for a sleeve resection and other procedures) or technical reasons (eg, adhesion); this was rarely needed for intra-operative bleeding.

Recovery phase

The recovery phase is also better for patients after VATS, compared with thoracotomy. The vital capacity was better initially and 1 year after VATS than with a thoracotomy, although the forced expiratory volume in one second, PaO_2, and partial pressure of carbon dioxide were not different [9,11]. The 6-minute walk was significantly better after VATS. Thoracotomy produced a significant impairment of vital capacity from 1 to 24 weeks after lobectomy ($P < .05$–0.001) [9]. Patients after a thoracotomy, compared with VATS, had significant impairment of the 6-minute walk 1 week after surgery ($P < .01$-0.001) [12].

Independence after VATS was greater than after a thoracotomy [9]. Transfer to care facilities or home nursing support was needed for 63% of open patients and only 20% of VATS patients ($P = .015$). VATS patients needed less personal care (10% versus 21%), wound or medical care (0% versus 13%), occupational or physical therapy (5% versus 13%), or other home support (5% versus 18%) than open patients [13].

A randomized trial from Germany showed fewer complications after VATS (14.2%) than thoracotomy (50%) [7]. A Japanese study showed cost (anesthesia charges, laboratory charges, and hospital charges) were less for the VATS approach [10]. Postoperative pain (visual pain scale, total dose of narcotic, need for additional narcotic, need for intercostal blocks, and sleep disturbances) is less after VATS than for thoracotomy [3].

The postoperative recovery appears to be better for the VATS approach than a thoracotomy. Demmy and Curtis [13] showed an earlier return to full preoperative activities, ($P < .01$) for the VATS patients. They also had better short-term and long-term quality of life (QOL) [14], less postoperative pain ($P = .014$) [3], and less shoulder dysfunction [15].

Overall, the QOL scores did not significantly differ between VATS and thoracotomy, but there are some QOL benefits to VATS [9]. Shoulder strength and range of motion was better at 1 week postoperatively, but the same at 3 months [15]. The incidence of limited activity at 3 months was greater after thoracotomy, and the return to full activities was faster with VATS [9].

VATS patients also had reduced postoperative release of both proinflammatory and anti-inflammatory cytokines. Although the postoperative release of tumor necrosis factor-α and interleukin (IL)-1β were minimal for both groups, the levels of IL-6, IL-8, and IL-10 were higher in the open group [16,17]. The clinical significance of these findings remains to be fully elucidated. For older patients, Demmy and Curtis [13] showed that VATS patients had less prolonged pain complaints (28% versus 56%, $P = .05$).

Postoperative complications in older patients undergoing anatomic pulmonary resection are common and contribute to prolonged hospitalization and associated health care costs. Compared with standard thoracotomy, successful use of a minimally invasive, non-rib spreading VATS approach for resection of early stage bronchogenic carcinoma is associated with reduced incidence and severity of postoperative complications in the elderly. A thoracoscopic approach to anatomic resection of early stage lung cancer may be preferred in these and possibly other high-risk patients. This issue warrants additional investigation.

Concerns unique to video-assisted thoracic surgery lobectomy

The biggest concerns regarding VATS lobectomy center on three issues: risk and management of intraoperative bleeding, tumor recurrence in the incision, and the adequacy of the cancer operation. The chances of these issues occurring appear to be small.

Some surgeons are concerned that dissection can be more difficult, so bleeding from the pulmonary artery can occur more easily and be more difficult to control by VATS than by thoracotomy [18]. However, several series have demonstrated that bleeding occurred in less than 1% of cases, so the risk of bleeding appears to be low for skilled VATS surgeons [1–16].

The oncologic impact of VATS lobectomy has been questioned with regards to recurrence in an incision, the completeness of the node dissection, the possible spread of cancer cells intraoperatively, and survival. Cancer rarely recurs in a thoracotomy incision, but fatal tumor recurrence in VATS incisions has been reported [19]. However, this happens in only 3 out of 1,321 (0.2%) cases [1–3]. Surgical technique is important to minimize this complication. Thoracic surgeons should not pull a tumor or an instrument that has touched a tumor through a small, unprotected incision. The author and colleagues have not experienced a trocar site recurrence in their VATS procedures for lung cancer since switching to the Lapsack for removal of the tumors [1]. As the skill of VATS surgeons has advanced, studies have shown that the number of lymph nodes removed by VATS was not inferior to that of thoracotomy. Systemic node dissection by VATS is technically feasible and safe, and seems acceptable for clinical stage I lung cancer [18].

Ultimately, the measure of any cancer treatment is survival. Another concern is about the adequacy of VATS lobectomy as a cancer operation. The true measure of any cancer treatment is survival. In a nonrandomized, prospective comparison of VATS versus open lobectomy for stage IA (T1N0) lung cancer, the 5-year survival was 90% and 85% for the VATS and thoracotomy groups, respectively ($P = .74$) [20]. Although some surgeons have reported exceptional survival (86%–94%) for stage I lung cancer after VATS lobectomy [21,22], others have reported the survival that is typically expected for surgical treatment of lung cancer (Table 2) [1–3]. It certainly appears that a VATS approach does not compromise the survival for lung cancer patients. Yamashita and colleagues [23] reported that carcinoembryonic antigen mRNA during video-assisted lobectomy was significantly higher than in subjects who underwent open lobectomy in a previous study (18 of 35 subjects; 51%; $P < .001$).

Key points: concerns

- Bleeding can occur with VATS lobectomy, but this appears to be rare and manageable.
- Recurrence in an incision can occur with VATS lobectomy, but this appears to be rare.
- Cure rates for a lobectomy by VATS and a thoracotomy appears to be the same.
- An anatomic lobectomy should be performed whether the procedure is performed by VATS or thoracotomy.

Table 2
Results of several reported series of video-assisted thoracic surgery lobectomy and pneumonectomy

VATS lobectomy/pneumonectomy				
Reference	No.	Cancer	Mortality	LOS
Lewis [21]	200	171	0	3.07
Yim [22]	214	168	1 (0.4%)	6.8
Kasada [20]	145	103	1 (0.8%)	NA
Walker [3]	159	131	3 (2%)	7.2
Roviaro [4]	169	142	1 (0.5%)	NA
Solaini [12]	112	99	0	5.8
McKenna [1]	1,100	935	8 (0.5%)	4.6
Daniels [2]	108	108	4 (4%)	3
Watanabe [25]	185	172	2 (1.1%)	N/A
Totals	2,390	2,029	22 (0.8%)	5.28

Abbreviations: LOS, length of stay in hospital; NA, not available.

Consensus statement regarding video-assisted thoracic surgery lobectomy

The consensus panel of the International Society of Minimally Invasive Cardiothoracic Surgery (ISMICS) made the following statements and recommendations regarding VATS for lobectomy in patients with clinical stage I non-small lung cancer [24]:

- VATS can be recommended to reduce the overall postoperative complications (class IIa, level A evidence).
- VATS can be recommended to reduce pain and overall improved functionality over the short term (class IIa, level B evidence).
- VATS can be recommended to improve the delivery of adjuvant chemotherapy delivery (class IIa, level B evidence).
- VATS can be recommended for lobectomy in clinical stage I and II non-small cell lung cancer patients, with no proven difference in stage specific survival, compared with open thoracotomy (class IIb, level B evidence).

Summary

Surgery remains the mainstay for the treatment of lung cancer. While pulmonary resection has been safe for years, there is a trend toward minimally invasive (VATS) pulmonary resections. Studies have now shown that standard complete cancer operations performed via VATS offer patients a shorter hospital stay and quicker recovery without compromising the cure rate for an operation performed via a thoracotomy.

References

[1] McKenna RJ Jr, Houck W, Fuller CB. Video-assisted thoracic surgery lobectomy: experience with 1,100 cases. Ann Thorac Surg 2006;81:421–6.

[2] Daniels LJ, Balderson SS, Onaitis MW, et al. Thoracoscopic lobectomy: a safe and effective strategy for patients with stage I lung cancer. Ann Thorac Surg 2002;74(3):860–4.

[3] Walker WS, Codispoti M, Soon SY, et al. Long-term outcomes following VATS lobectomy for non-small cell bronchogenic carcinoma. Eur J Cardiothorac Surg 2003;23(3):397–402.

[4] Roviaro G, Varoli F, Vergani C, et al. Long term survival after video-assisted thoracoscopic surgery lobectomy for stage I lung cancer. Ann Thorac Cardiovasc Surg 2003;9(1):14–21.

[5] Kirby TJ, Mack MJ, Landreneau RJ, et al. Initial experience with video-assisted thoracoscopic lobectomy. Ann Thorac Surg 1993;56:1248–53.

[6] Giudicelli R, Thomas P, Lonjon T, et al. Major pulmonary resection by video assisted mini-thoracotomy. Initial experience in 35 patients. Eur J Cardiothorac Surg 1994;8:254–8.

[7] Hoksch B, Ablassmaier B, Walter M, et al. Complication rate after thoracoscopic and conventional lobectomy. Zentralbl Chir 2003;128(2):106–10.

[8] Koizumi K, Haraguchi S, Hirata T, et al. Lobectomy by video-assisted thoracic surgery for lung cancer patients aged 80 years or more. Ann Thorac Cardiovasc Surg 2003;9(1):14–21.

[9] Cheng D, Downey RJ, Kernstine K, et al. Video assisted thoracic surgery in lung cancer resection. Innovations 2007;2(6):261–92.

[10] Nakajima J, Takamoto S, Kohno T, et al. Costs of videothoracoscopic surgery versus open resection for patients with of lung carcinoma. Cancer 2000;89(Suppl 11):2497–501.

[11] Kaseda S, Aoki T, Hangai N, et al. Better pulmonary function and prognosis with video-assisted thoracic surgery than with thoracotomy. Ann Thorac Surg 2000;70:1644–6.

[12] Solaini L, Prusciano F, Bagioni P, et al. Video-assisted thoracic surgery major pulmonary resections. Present experience. European Journal of Cardio-Thoracic Surgery 2001;20(3):437–42.

[13] Demmy TL, Curtis JJ. Minimally invasive lobectomy directed toward frail and high-risk patients: a case-control study. Ann Thorac Surg 1999;68(1):194–200.

[14] Sugiura H, Morikawa T, Kaji M, et al. Long-term benefits for the quality of life after video-assisted thoracoscopic lobectomy in patients with lung cancer. Surg Laparosc Endosc Percutan Tech 1999;9(6):403–10.

[15] Li WW, Lee RL, Lee TW, et al. The impact of thoracic surgical access on early shoulder function video-assisted thoracic surgery versus posterolateral thoracotomy. Eur J Cardiothorac Surg 2003;23(3):390–6, IL 28.

[16] Yim AP, Wan S, Lee TW, et al. VATS lobectomy reduces cytokine responses compared with conventional surgery. Ann Thorac Surg 2000;70(1):243–7.

[17] Leaver HA, Craig SR, Yap PL, et al. Phagocyte activation after minimally invasive and conventional pulmonary lobectomy. Eur J Clin Invest 1996;26(Suppl 1):230–8.

[18] Sugi K, Sudoh M, Hirazawa K, et al. Intrathoracic bleeding during video-assisted thoracoscopic lobectomy and segmentectomy. Japanese Journal of Thoracic Surgery 2003;56(11):928–31.

[19] Downey RJ, McCormack P, LoCicero J, et al. Dissemination of malignant tumors after video-assisted thoracic surgery: a report of twenty-one cases. J Thorac Cardiovasc Surg 1996;111:954–61.

[20] Kaseda S, Aoki T. Video-assisted thoracic surgical lobectomy in conjunction with lymphadenectomy for lung cancer. J Jap Surg Soc 2002;103(10):717–21.

[21] Lewis RJ, Caccavale RJ. Video-assisted thoracic surgical non-rib spreading simultaneously stapled lobectomy (VATS(n)SSL). Semin Thorac Cardiovasc Surg 1998;10:332–7.

[22] Yim APC, Ko KM, Chau WS, et al. Thoracoscopic major lung resections: an Asian perspective. Semin Thorac Cardiovasc Surg 1998;10:326–31.

[23] Yamashita JI, Kurusu Y, Fujino N, et al. Detection of circulating tumor cells in patients with non-small cell lung cancer undergoing lobectomy by video-assisted thoracic surgery: a potential hazard for intraoperative hematogenous tumor cell dissemination. J Thorac Cardiovasc Surg 2000;119(5):899–905.

[24] Nakajima Downey RJ, Cheng D, Kernstine K, et al. Video assisted thoracic surgery in lung cancer resection: a consensus statement of the International Society of Minimally Invasive Cardiothoracic Surgery. Innovations 2007;2(6):293–302.

[25] Watanabe A, Osawa H, Watanabe T, et al. Complications of major lung resections by video-assisted thoracoscopic surgery. Japanese Journal of Thoracic Surgery 2003;56(11):943–8.

ELSEVIER
SAUNDERS

Thorac Surg Clin 18 (2008) 281–287

THORACIC
SURGERY
CLINICS

The VATS Lobectomist: Analysis of Costs and Alterations in the Traditional Surgical Working Pattern in the Modern Surgical Unit

William S. Walker, MA, FRCS, FRCSE*,
Gianluca Casali, MD

Department of Thoracic Surgery, Royal Infirmary of Edinburgh, University of Edinburgh, Little France, Edinburgh EH16 4SA, Scotland, UK

Video-assisted thoracic surgery (VATS) lobectomy remains an infrequently performed procedure. Despite a noticeable increase in interest in VATS resection in both the United States and Japan, uptake in Europe has changed little in a decade. Various factors account for both these differences and the slow growth in the field overall. Among these remain concerns as to oncologic validity, especially regarding adequacy of mediastinal node management; safety; training; cost issues; and the exact fit for VATS techniques within the profile of a thoracic service. These latter two issues form the basis for this article.

Assessment of the comparative costs for VATS and open lobectomy is complicated by several issues. These include the differing medical and social conditions in various countries; the operative techniques used for both VATS and open resection; and the inability of current cost models to capture aspects of quality of life, particularly in a generally elderly patient group. To address this question this article reviews the consideration likely to arise from variations in VATS technique and geographic setting; considers actual costs encountered within a Northern European model of care; and, working from these data, determines the extent of any financial difference. Based on these findings, also reviewed are the variations that might be anticipated within other care systems and using different VATS techniques.

* Corresponding author.
E-mail address: wsw@holyrood.ed.ac.uk (W.S. Walker).

Incorporation of a VATS major pulmonary resection program within the structure of a general thoracic surgery unit places significant emphasis on selection, work-up, and integration. VATS techniques are but one element in the armamentarium available to the thoracic surgeon. The skill lies as much in making the correct choice on behalf of the patient as in the conduct of the procedure itself.

Analysis of costs

Models of care

Across the world care delivery and cost attribution varies widely. Models of care tend to fall into several different patterns (Table 1). All are heavily interested in cost containment but the drivers for this and the methods for achieving it vary widely. For example, in a charity or state-funded program, restricting physician fees is not relevant, whereas in a market-driven process this is one of the more vulnerable areas within the cost envelope. Conversely, non-business–orientated programs are very sensitive to disposable and equipment costs because there is no easy method of passing these costs back to the funding source by way of individualized patient charges or agreed care package prices.

Different care model environments focus on different aspects of the overall cost and make it difficult to generalize between systems regarding the relative merits of one procedure over the other. It is clear, however, that the level of patient

1547-4127/08/$ - see front matter
doi:10.1016/j.thorsurg.2008.06.003

Table 1
Variation in models of care

Funding agent	Drivers for economy	Major economy areas
Self payment	Acceptable end user cost Package deal envelope Provider profit	Physician fees, procedure cost, bed stay
Insurer	Premium competition Provider profit	Physician fees, procedure cost, bed stay
National health program	Budget compliance, meet service demand	Equipment, disposables, bed stay
Charity or aid	Maximize number of benefitted patients	Equipment, disposables

choice is quite different within these models varying from complete for self-funded patients to effectively nil for those relying on charitable funding. Such issues as cosmesis, pain, and functionality can be of importance in one group but of little relevance to another group. Similarly, state-funded programs, although unsympathetic in general terms to innovation for fear of the cost implications, are sensitive to societal norms and expectations regarding inpatient stay. Inpatient stay becomes an issue specific to the health care environment and not meaningfully comparable between systems. For example, based on contemporary reports, a postoperative stay of 7 to 15 days following major pulmonary resection is acceptable in Northern Europe and in Japan 21 days is not unusual. In the United States, however, postoperative inpatient stay is typically a matter of a few days. Or is it? Many institutions "discharge" to adjacent hotels, sometimes even physically connected to the hospital. Is this really discharge or simply transfer of a portion of the cost burden? Furthermore, notions of fitness for discharge vary widely. In general terms, within a northern European environment, discharge with a Heimlich-type valve is relatively unusual and home care issues often delay discharge while domestic circumstances are assessed and, if necessary, modified to allow supported discharge. Conversely, in other care environments there may be a strong financial imperative for the patient or the family to accept a high level of ongoing care requirement at home. Finally, the patient substrate is likely not directly comparable between these systems. Quoted package deals, for example, require a higher degree of certainty than may be afforded by elderly or multiply comorbid patients.

Operative techniques

What is a "VATS" lobectomy? This question is addressed elsewhere in this issue, but does require some consideration here to provide a background for discussion. The three broad technique

Table 2
Video-assisted thoracic surgery lobectomy techniques

Technique	Outline	Use of stapling devices	Special instrumentation
Simultaneously stapled lobectomy [1]	Endoscopic procedure Stapled division of fissures Mass stapling of hilar pedicle	Endo-stapler for fissure Open stapler for pedicle	Optional, not essential
Minithorcotomy [2,6]	Per-incisional surgery Direct vision Orthodox dissection Supplementary light and vision by videothoracoscope	Optional, not essential Stapler often used for bronchus	Optional, not essential
Endoscopic hilardissection [3–5,18,21]	Endoscopic procedure Anatomic hilar dissection Individual structure division	Endo-staplers and clip devices used extensively	Endo-dissectors and shears usually desirable

groupings are described in Table 2. There are clearly multiple variations in the exact implementation of these operative strategies and the simultaneously stapled technique [1] is probably now significantly less used than the other options. Of the other two, it can be observed that the minithoracotomy technique [2] could require the least investment in endoscopic instrumentation and is an automatically cheaper approach. Whether this technique truly represents a VATS lobectomy or the ultimate development of a muscle-sparing approach is a matter for debate but it does represent a minimally invasive approach and is widely used, particularly in Asia, perhaps reflecting in part the potential for minimal disposables costs. Endoscopic hilar dissection [3–5], a true endoscopic analog of open resection, requires the greatest investment in training and disposables but may be gaining in overall acceptance.

One might reasonably also ask, however, "what is an open lobectomy," for there are significant variations among surgeons regarding incision size, rib division or resection or cartilage separation in gaining entry, use of stapling devices, and lymph node management, to list merely the obvious operative issues. These differences potentially impact on operative time, postoperative recovery, and discomfort. Despite the close scrutiny of VATS procedures both in terms of cost and long-term outcome, knowledge of the outcomes associated with various forms of thoracotomy and styles of resection remains quite uncertain.

Cost analysis

Few authors [6–10] have specifically considered the economic aspects of VATS lobectomy, although such factors as inpatient stay have been described in comparative series. Such analyses have to be reviewed within the context of the health care system and region of origin, and one criticism that may be leveled at almost all the available studies is that of inadequate sample sizes.

In an early study, Giudicelli and coworkers [6] noted a trend toward reduced mean in patient stay in patients undergoing VATS lobectomy using a minithoracotomy technique compared with a parallel control group undergoing muscle-sparing thoracotomy (days: 12 VATS versus 15 open). In an otherwise negative randomized study comparing VATS lobectomy using an endoscopic technique with open resection by a muscle-sparing thoracotomy, Kirkby and colleagues [7] found a reduced (but not significant) inpatient stay with VATS lobectomy (7.1 versus 8.3 days) associated with significantly ($P < .05$) fewer complications in the VATS cases. Lewis and colleagues [8] in a retrospective cost comparison study comparing 15 open and 15 VATS patients found open resection to cost almost twice as much as a VATS procedure using his simultaneously stapled technique reflecting reduced advanced care usage and shorter inpatient stay. These data were not significant, however, because of the small sample size. Sugi and colleagues [9], in a Japanese setting, compared 10 VATS lobectomies with 20 open thoracotomy controls. They found an increase in surgical disposables of over $3000 with a VATS procedure and little benefit with regard to postoperative stay, which averaged 25.2 days for the VATS groups as opposed to 27.7 for the open thoracotomy group. Nakajima and colleagues [10] presented a mixed study, also from Japan, involving early stage bronchogenic carcinoma and metastases both resected by lobectomy or sublobar excision using both VATS and open surgery. Included were 66 open thoracotomies and 36 VATS procedures. Overall hospital charges were less with VATS resection and mean inpatient stay was significantly reduced at 17.3 days compared with 23.8 for open resection. Interestingly, Liu and colleagues [11] reported a mixed VATS series of 2300 patients operated on over a 7-year period and commented that cost effectiveness was a vital consideration for the survival of VATS techniques in an Asian setting. Key to this was the use of conventional instruments rather than expensive endoscopic disposable devices.

The authors recently undertook a detailed comparative analysis of costs associated with VATS and open resection in their unit. This study clearly reflects United Kingdom costs and the economics of a nationalized health care system. Nonetheless, there are some interesting findings.

This analysis, reported in preliminary form [12] and presented in outline here, reviewed cost and outcome data for 346 patients undergoing pulmonary lobectomy between January 2004 and December 2006. Most resections were for early stage lung cancer (stage I or II), with the remainder a mix of metastatic disease and inflammatory or benign conditions. A total of 93 VATS lobectomies and 253 open thoracotomy resections were performed. In the VATS group 47% of patients had an upper lobectomy versus 52% in the open group (P = ns). Direct medical costs (ie, disposables, theater time, high dependency unit stay, and hospital) were determined and stratified by lobectomy type.

It was found that the mean theater cost for a VATS lobectomy was significantly higher at 2533 ± 230€ versus 1280 ± 54€ for a thoracotomy lobectomy ($P = .00001$). The mean high dependency unit cost, however, was reduced in VATS cases at 1713 ± 236€ compared with 2571 ± 80€ for a thoracotomy lobectomy ($P = .00001$) reflecting reduced bed usage in the VATS group. Similarly, the mean cost of the postoperative inpatient ward stay was less in the VATS lobectomy group at 3776 ± 281€ as against 4325 ± 154€ for open resection ($P = .00001$).

Overall, the cost of a VATS lobectomy was similar to that of an open lobectomy. Single-lobe resection worked out at 155€ cheaper by VATS (8023 ± 565€ VATS versus 8178 ± 167€ open; $P = .0002$) and bilobectomy was 47€ more by VATS (8702 ± 350€ VATS versus 8655 ± 466€ open; P = ns). Bed stay for VATS patients was 5.54 ± 0.37 days and 6.87 ± 0.19 days for open thoracotomy patients ($P = .001$). The authors reviewed a broad panel of postoperative complications and found a consistent trend for events, notably chest infection and lobar-sublobar collapse, to be more frequent in the thoracotomy group than in the VATS group (54.2% versus 46.2%; $P = 0.19$).

This study demonstrates that VATS lobectomy does not present a financial challenge to the parent institution. It does, however, offer reduced postoperative complication rates and an easier recovery path with a consequent decrease in bed occupancy. The authors did not model the potential for income generation provided by the liberated bed days because this would not be of direct relevance within a nationalized health care system, but this reduced bed stay should have an opportunity value for more treated patients within a more market-sensitive environment.

Such studies are confounded by various factors: the lack of randomized data; sample volumes; and, above all, the enormous differences in medical and social environments across the world. As observed by Van Schil [13], there is a definite need for high-quality prospective randomized studies in this area as is true of most comparisons between VATS and open surgery. Whether such an ideal is attainable is another matter. Patients are understandably reluctant to agree to a randomization process that potentially allocates them to a more invasive procedure. Clinicians have to rely on parallel cohort data, which generate major issues regarding bias. Nonetheless, there seems little prospect of true randomized studies and one must hope that the use of large collaborative databases analyzed using appropriate matching tools, such as propensity scoring, will provide robust answers acceptable to all concerned.

Alterations in the traditional surgical working pattern in the modern surgical unit

Where does VATS lobectomy fit within the procedural base of a busy general thoracic surgery unit? To answer this question one must first consider fundamental principles and from that deduce where VATS may be relevant.

As a first observation, the use of VATS is, in part, a function of the skill and aptitude of the surgeon. It is a core consideration that any use of VATS techniques should not be to the prejudice of the patient simply to achieve a "modernist" approach. There are established pathways for diagnosis, assessment, staging, and management that function perfectly well without the use of VATS. Clinicians must aim to avoid compromise and to look toward those areas where the use of VATS techniques may enhance the standard of care delivery to their patients. VATS lobectomy is simply the apex of a pyramid of VATS interventions. In the authors' unit, the philosophy is one of integration (Fig. 1), so that VATS plays a greater or lesser role in the various stages of patient management. It is also the case that just as VATS is but one of the available tools, so the possibility must always be considered whether it is better to select a different approach, integrate both open and VATS techniques, or indeed change to an open procedure.

Staging and assessment

Before considering VATS lobectomy in detail it is important to recognize the major contribution

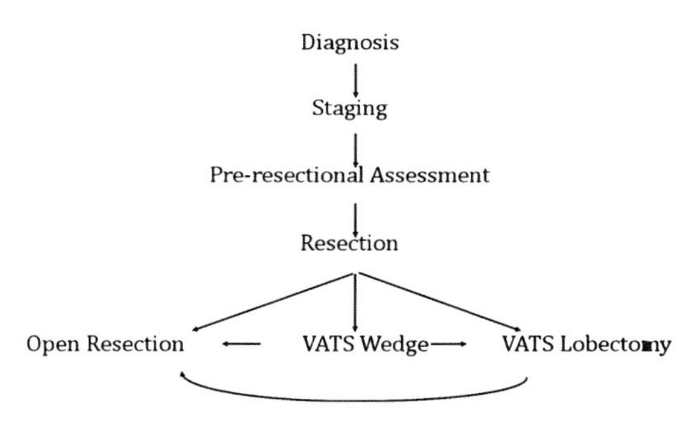

Fig. 1. Stylized flow sequence for oncologic assessment and management of bronchogenic carcinoma.

to patient welfare afforded by VATS staging and assessment. In the United Kingdom approximately 6% of thoracotomies are in the "open and shut" category. It may be argued that this is an acceptable figure if all patients are to be given the potential advantage of resection where preoperative data are in the balance. In the authors' experience, however, the use of preliminary VATS inspection before any intended resection has cut the rate of no resection thoracotomy to less than 1% consistent with the findings of others [2,14,15]. They advocate routine insertion of a videothoracoscope in all potential major pulmonary resection cases. This takes but a few minutes and can help in several important ways. Pleural or other unexpected metastatic disease can be excluded. Obvious invasion of structures that makes surgery impossible can be identified. On occasion, irresectability implied on preoperative scans can be refuted providing the opposing pathway. Also, and often not considered, VATS assessment provides the best opportunity to make a patient-based judgment of relative resectability. Consider, for example, the patient with marginal pulmonary function in whom a lobectomy is feasible but a pneumonectomy ill advised. VATS assessment can confirm the situation before the surgeon commits to thoracotomy. Similarly, chest wall or aortic invasion might be operable in a good-risk candidate but inappropriate in other cases. Whether the adverse finding negating thoracotomy is absolute or relative, the option to proceed to alternative oncologic therapies is hugely enhanced by avoiding the major trauma of an unnecessary thoracotomy. Recent work suggests that in some instances a single-port approach can provide adequate assessment [16]. As a final consideration, routine assessment of cancer cases provides familiarity for the surgeon and team in this field and consequently a sound basis for judgment in undertaking a VATS lobectomy. Better this gradated approach than the abrupt transition from relatively minor VATS interventions to a major endoscopic resection.

Video-assisted thoracic surgery lobectomy

The role of VATS lobectomy in a thoracic unit is subject to several interdependent issues. These include surgical (and other operating room team member) competencies; unit case volume; core surgical philosophy; and the staging strategy used. Beyond these considerations program size may also be governed by economic considerations.

Program size

It is relatively easy to calculate the maximum reasonable percentage of VATS resections within a unit. Specimen extraction limits tumor size in any practical sense to about 5 cm and the requirement for a clean hilar dissection area makes selection of more peripheral lesions desirable. T1 and T2 nonhilar lesions are preferred. Lymph node management is a matter for debate. In the authors' experience, using the endoscopic VATS lobectomy technique complete hilar dissection is routine and mediastinal adenectomy on the right is perfectly feasible but it is difficult to clear stations 7 and L2 and L4 on the left. This may be easier to undertake using a minithoracotomy technique. Arising from these considerations, the authors have opted for extensive preoperative mediastinal assessment, which currently includes CT, CT with positron emission tomography, and routine mediastinoscopy in all cases with inspection and sampling of ipsilateral stations 2 and 4, station 7, and contralateral 4 [17]. The intent is to limit VATS resection to stage I and II disease. In their practice more advanced cases are the appropriate candidates for open surgery. On this basis, approximately 30% to 40% of resections undertaken could be candidates for VATS lobectomy. Clearly, if a unit had a particular interest in resection of advanced-stage cases, the proportion is reduced. It is perfectly possible to advance into N2 category cases subject to the previously mentioned considerations and into hilar disease or larger tumors. In these instances there may be issues regarding the required size of the utility port and of rib retraction being required. Nonetheless, there can be a strong argument for proceeding in this direction on the basis that these patients can progress to adjuvant therapy more easily than after standard thoracotomy [18]. Where to draw the line in this process is a value judgment for the unit concerned but likely only a high-volume program generates the necessary skills to support resection of advanced cancer by VATS lobectomy.

Competency

As with all forms of surgery, maintenance of skill requires that a certain minimum number of VATS lobectomy should be undertaken each year. In the authors' opinion, a minimum of 20 cases per year should be undertaken by a VATS lobectomist, ideally more. Logically, unit structure should provide for two surgeons in a group to have this subspecialization, whereas the others focus on different aspects. A program of 40 to 50 VATS lobectomies supports two surgeons within a likely total case volume of 130 to 150 resections per year.

There are as yet no specific credentialing in VATS lobectomy, with some arguing that this is another skill that the contemporary trainee acquires in due course. This is not the case. Data [19] suggest that a learning curve of 50 cases is required to reach full competence. It follows that units wishing to develop this aspect of thoracic surgery require investing in individuals with a proved track record in the field and developing talent within their own group. A fully trained surgeon is necessary to ensure that patient care is not compromised with respect to open surgery.

Where specific training is not a possibility, the option to develop a de novo program exists but is a long and responsible road. In this circumstance one advises a background of good endoscopic skills, considerable experience with open resection, and a willingness to progress in a slow and incremental manner. Accordingly, initial VATS assessment might progress to limited preliminary mobilization followed in later cases by mobilization of structures until at some point favorable fissures and clear anatomy facilitate a successful lobectomy. In some instances, it may be found helpful to start with a minithoracotomy approach and to migrate from that to an endoscopic procedure. Whichever strategy is adopted, however, one should be aware that this is a difficult way forward and it is better to learn from the experience of others. To paraphrase Sir William Osler, learning in such a manner "is to sail an uncharted sea."

Competency issues are not restricted to the surgeon. It is pointless embarking on a VATS program unless nursing and anesthesia colleagues are prepared to support this process and to accept the additional requirements of VATS lobectomy and increased operating time necessary. Time pressure makes for bad surgery.

Summary

It is beyond the scope of this article to review the advantages of VATS lobectomy, but the data in support of this technique are increasing progressively. There is excellent evidence to support the oncologic equivalence and safety profile as compared with open thoracotomy [4,5,20,21], and data that demonstrate the reduced pain associated with VATS resection [22,23]. Also, reduction in immune disturbance provides a tantalizing glimpse of one additional potential modality of benefit for less traumatic surgery [24]. Unfortunately, in the economic world, equivalence, preferably with less cost, is the test applied. Whatever the societal benefit of improved quality of life following surgery, this has no cost benefit attached. From the foregoing discussion one can conclude that VATS lobectomy is no more costly than open resection and does generate additional hospital beds. The authors remain uncertain as to the preferred form of VATS lobectomy but it seems that the reduced trauma of the endoscopic procedure is associated with more benefit in terms of shorter hospitalization albeit at the cost of some increase in operating time [25]. VATS techniques and lobectomy sit comfortably within the structure of any thoracic unit requiring little adjustment to established process. It is likely that ultimately 30% or thereabouts of major pulmonary resection will be undertaken using this technique and that VATS interventions will aid patient assessment regardless of stage or ultimate intended therapy. Competency and responsible use remain paramount considerations.

References

[1] Lewis RJ, Caccavale RJ, Sisler GE, et al. One hundred video-assisted thoracic surgical simultaneously stapled lobectomies without rib spreading. Ann Thorac Surg 1997;63:1415–21.

[2] Giudicelli R, Thomas P, Lonjohn T, et al. Major pulmonary resection by video assisted mini-thoracotomy. Eur J Cardiothorac Surg 1994;8:254–8.

[3] Roviaro G, Varoli F, Nucca O, et al. Long term outcomes after videothoracoscopic resection for stage I lung cancer. Chest 2004;126:725–32.

[4] Walker WS, Codispoti M, Soon SY, et al. Long-term outcomes following VATS lobectomy for non-small cell bronchogenic carcinoma. Eur J Cardiothorac Surg 2003;23(3):397–402.

[5] McKenna RJ Jr, Houck W, Fuller CB. Video-assisted thoracic surgery lobectomy: experience with 1,100 cases. Ann Thorac Surg 2006;81(2):421–5 [discussion 425–6].

[6] Giudicelli R, Thomas P, LonjonT, et al. Video-assisted minithoracotomy versus muscle-sparing thoracotomy for performing lobectomy. Ann Thorac Surg 1994;58(3):712–7 [discussion 717–8].

[7] Kirkby TJ, Mack MJ, Landreneau RJ, et al. Lobectomy–video-assisted thoracic surgery versus muscle-sparing thoracotomy: a randomized trial. J Thorac Cardiovasc Surg 1995;109:997–1002.

[8] Lewis RJ, Caccavale RJ, Sisler GE, et al. Is video-assisted thoracic surgery cost effective? N J Med 1996;93(12):35–41.
[9] Sugi K, Kaneda Y, Nawata K, et al. Cost analysis for thoracoscopy: thoracoscopic wedge resection and lobectomy. Surg Today 1998;28(1):41–5.
[10] Nakajima J, Takamoto S, Kohno T, et al. Costs of videothoracoscopic surgery versus open resection for patients with of lung carcinoma. Cancer 2000; 89(Suppl 11):2497–501.
[11] Liu HP, Wu YC, Liu YH, et al. Cost-effective approach of video-assisted thoracic surgery: 7 years experience. Chang Gung Med J 2000; 23(7):405–12.
[12] Casali G, Walker W. VATS lobectomy: can we afford it? Inter J Cardiovascular and Thor Surg 2007;6(Suppl 2):S191.
[13] Van Schil P. Cost analysis of video-assisted thoracic surgery versus thoracotomy: critical review. Eur Respir J 2003;22:735–8.
[14] Loscertales J, Jimenez-Merchan R, Arenas-Linares C, et al. The use of videoassisted thoracic surgery in lung cancer: evaluation of respectability in 296 patients and 71 pulmonary exeresis with radical lymphadenectomy. Eur J Cardiothorac Surg 1997;12: 892–7.
[15] Sonett JR, Krasna MJ. Thoracoscopic staging for intrathoracic malignancy. In: Yim APC, Hazelrigg SR, Izzat MB, et al, editors. Minimal access cardiothoracic surgery. Philadelphia: WB Saunders; 2000. p. 183–93.
[16] Rocco G, Brunelli A, Jutley R, et al. Uniportal VATS for mediastinal nodal diagnosis and staging. Inter J Cardiovascular and Thor Surg 2006;5: 430–2.
[17] De Leyn P, Lardinois D, Van Schil P, et al. ESTS guidelines for preoperative lymph node staging for non small cell lung cancer. Eur J Cardiothorac Surg 2007;32:1–8.
[18] Petersen RP, Pham D, Burfeind WR, et al. Thoracoscopic lobectomy facilitates the delivery of chemotherapy after resection for lung cancer. Ann Thorac Surg 2007;83(4):1245–9 [discussion 1250].
[19] Ferguson J, Walker W. Developing a VATS lobectomy programme: can VATS lobectomy be taught? Eur J Cardiothorac Surg 2006;29(5):806–9.
[20] Thomas P, Doddoli C, Yena S, et al. VATS is an adequate oncological operation for stage I non-small cell lung cancer. Eur J Cardiothorac Surg 2002; 21(6):1094–9.
[21] Gharagozloo F, Tempesta B, Margolis M, et al. Video-assisted thoracic surgery lobectomy for stage I lung cancer. Ann Thorac Surg 2003;76(4): 1009–14 [discussion 1014–5].
[22] Walker WS. Video-assisted thoracic surgery (VATS) lobectomy: the Edinburgh experience. Semin Thorac Cardiovasc Surg 1998;10:291–9.
[23] Ohbuchi T, Morikawa T, Takeuchi E, et al. Lobectomy: video-assisted thoracic surgery versus posterolateral thoracotomy. Jpn J Thorac Cardiovasc Surg 1998;46(6):519–22.
[24] Walker WS, Leaver HA. Immunologic and stress responses following video-assisted thoracic surgery and open pulmonary lobectomy in early stage lung cancer. Thorac Surg Clin 2007;17:241–9.
[25] Shigemura N, Akashi A, Funaki S, et al. Long-term outcomes after a variety of video-assisted thoracoscopic lobectomy approaches for clinical stage IA lung cancer: a multi-institutional study. J Thorac Cardiovasc Surg 2006;132(3):507–12.

ELSEVIER
SAUNDERS

Thorac Surg Clin 18 (2008) 289–295

THORACIC
SURGERY
CLINICS

Robotically Assisted Lobectomy: Learning Curve and Complications

Franca M.A. Melfi, MD*, Alfredo Mussi, MD

Cardiac and Thoracic Department, University of Pisa, Via Paradisa\2, 56100 Pisa, Italy

There are different types of robotic devices, some as simple as camera voice-control [1,2]. Today, the da Vinci Robotic System (Surgical Intuitive, Mountain View, California) is the only complete surgical system applied in surgical practice. This system includes a master remote console, a computer controller, and a three to four arm surgical manipulator with fixed remote center kinematics connected by electrical cables and optic fibers (Fig. 1).

The master console is connected to a surgical manipulator with two to three instrument arms and a central arm to guide the endoscope. Two master handles at the surgeon's console are manipulated by the user. The position and the orientation of the surgeon's hands on the handles trigger highly sensitive motion sensors, which transfer the surgeon's movements to the tip of the instrument at a remote location.

The surgical arm cart provides three degrees of freedom: (1) pitch, (2) yaw, and (3) insertion. Attached to the robot arm is the surgical instrument, the tip of which is provided with a mechanical cable-driven wrist. This adds four more degrees of freedom: (1) internal pitch, (2) internal yaw, (3) rotation, and (4) grip. To increase precision, the system uses downscaling from the motion of the handles to that of the surgical arms. In addition, unintended movements caused by human tremor are filtered by a 6-Hz motion filter.

This system overcomes many of the obstacles of thoracoscopic surgery (Table 1). It increases dexterity, restores proper hand-eye coordination and ergonomic position, and improves visualization. In addition, it renders now feasible surgical operations that were technically difficult or even impossible. Instruments with increased degrees of freedom greatly enhance the surgeon's ability to manipulate the tissues. The system is designed so that the surgeon's tremor can be compensated on the end-effector motion through appropriate hardware and software filters. Moreover, this robotic system eliminates the fulcrum effect, making instrument manipulation more intuitive. By most accounts, the enhanced vision afforded by these systems is remarkable. The three-dimensional view with depth perception is a marked improvement over the conventional thoracoscopic camera views. A further advantage is the surgeon's ability directly to control a stable visual field with increased magnification and maneuverability. All of this creates images with increased resolution that, combined with the increased degrees of freedom and enhanced dexterity, greatly enhances the surgeon's ability to identify and dissect anatomic structures and to construct microanastomoses.

Nevertheless, there are several disadvantages to the da Vinci Systems. Robotic surgery is a new technology and its uses and efficacy have not yet been well established. Another disadvantage is the size of this system. This is an important disadvantage in today's already crowded operating rooms. Moreover, one of the potential disadvantages identified is a lack of compatible instruments and equipment. Lack of certain instruments increases reliance on tableside assistants to perform part of the surgical operation. This, however, is a transient disadvantage because new technologies have been and will continue to be developed to address these shortcomings. Most of these disadvantages will be remedied with time and improvements in technology.

* Corresponding author.

E-mail address: f.melfi@med.unipi.it (F.M.A. Melfi).

1547-4127/08/$ - see front matter
doi:10.1016/j.thorsurg.2008.06.001

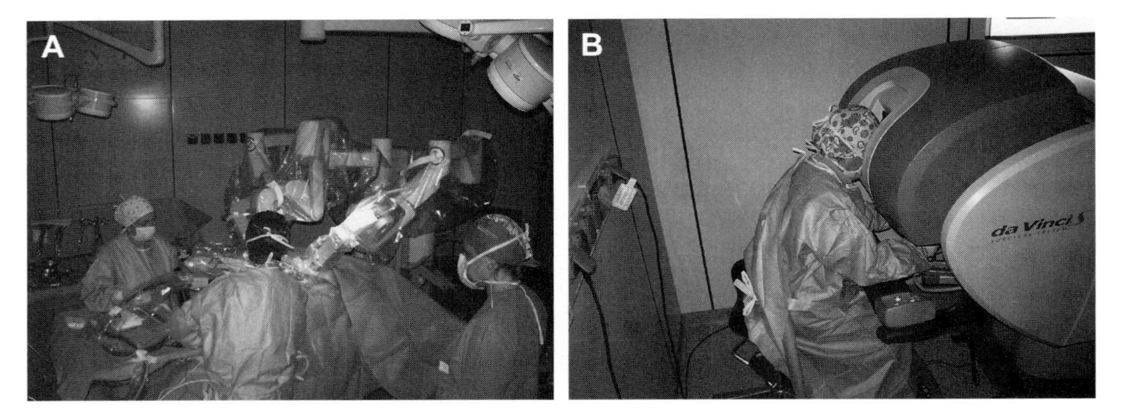

Fig. 1. (*A*) The surgical team. (*B*) The surgeon at master remote console.

Training of surgical team: learning curve

To perform robotic surgery in a safe and straightforward manner, it is necessary to standardize procedures and establish operative schemes. This robotic device requires meticulous preparation in terms of set-up of the system and its placement at the operating table.

The main body of the machine (surgical cart) and the robotic arms must be placed in relation to the side of the lesion. Only when the robotic cart has been positioned appropriately and the patient placed in the chosen position, can the robotic arms be brought into the operative field. The transition from traditional surgery to advanced totally robotic surgery is not immediate. Just as in the passage from open surgery to minimally invasive technique, here too certain precise organizational and didactic routes must be followed. Advanced training on robotic systems provides the surgeon and scrub nurses (surgical team) confidence when operating in tiny intracavitary spaces. This allows the surgical team to activate and maintain the entire operative system, recognize and correct errors, and take charge of handling all materials and instruments. Approximately, 20 days are necessary to complete adequate training.

To expand these new capabilities, robotic surgical procedures were developed and objective-based curriculum levels were designed to optimize surgeon and team training, which is necessary to obtain the best early clinical results (Box 1).

Binocular and three-dimensional vision, restricted operative field, handling of joysticks, robotic surgical instruments, movements of robotic arms and surgical instruments, and absence of tactile feedback are important aspects that the surgeon needs to become familiar with

Table 1
Advantages and disadvantages of videothoracoscopic surgery versus robotic surgery

	Videothoracoscopic surgery	Robotic surgery
Advantages	Well-developed technology	Three-dimensional imaging
	Affordable and ubiquitous	Dexterity
	Proved efficacy	Seven degrees of freedom
		No fulcrum effect
		No physiologic tremors
		Scale motions
		Ergonomic position
Disadvantages	Loss of touch sensation	No tactile feedback
	Compromised dexterity	Expensive
	Limited degrees of motion	Unproved benefit
	Fulcrum effect	
	Amplification of physiologic tremors	

Box 1. The scheme for the learning curve

Nurse team

1. Setup robotic system
   Connection of the console to the robotic cart (electric cables and optic fibers)
   Self-test (3 minutes): the test system undergoes an automatic check-up
   Draping the robotic arms in sterile nylon covers
2. Setup optic system
   Frontal or inclined position of the scope (0 degrees or 30 degrees)
   White balancing
   Setting of two- or three-dimensional

Surgeon's team

1. Trocar placement
   The best positioning of the trocars is established in relation to the side of the lesion to have an excellent, unobstructed view of the chest cavity without arm impingement and interference
2. Patient positioning
3. Robot cart positioning
4. Surgical procedure (pneumothorax, mediastinum lesions)

during the learning curve. After an initial theoretic phase, training at the console represents the surgeon's first real impact with robotic surgery. This can be performed using either mechanical or animal models: in this regard, it is auspicable to have a special robot available that can be used exclusively for teaching purposes. The most frequent procedure used in the initial phase of the learning curve is the treatment of pneumothorax and mediastinum lesions (neurinoma, pericardial cystis). These procedures represent an ideal training model, because they provide the means for learning basic procedures combined with a relatively simple technique. As surgical experience grows, indications for robotic technique can be extended to include an increasing number of procedures while obtaining results that conform to specific standards. Improvements in the quality of surgery and reductions in operative times correspond to standardization of procedure in terms of patient positioning, trocar placement, and consequent reduction time of the surgical cart and mechanical arms placement in the surgical field. A minimum of 20 procedures is necessary for the learning curve for both teams. In addition, the active cooperation with anesthesiologists is auspicious, mostly with regards to patient and system positioning so as not to hamper patient monitoring.

Robotic lobectomy: technical aspects

Many of the robotic procedures can be performed by a single operator. This is true of procedures that require few accurate maneuvers (dissection or coagulation) in a restricted and well-defined field: enucleation of condroma, excisions of mediastinal masses, thymectomies. During major resections, such as lobectomies, however, some maneuvers must necessarily be performed by the assistant surgeon, given the need of a fourth arm. Maintaining a correct position with appropriate tension of the lung parenchyma is top priority for identifying and dissecting the hilum structures (vessels and bronchus), as is suction, passing the sutures in the chest cavity, and appropriate positioning of the stapler. To perform these maneuvers the role of the assistant surgeon is mandatory, and they must always be at hand at the operating table. The new da Vinci System (da Vinci S), however, overcame most of these technical difficulties, thanks to multiquadrant access of a fourth arm (the robotic arms have greater range of motion and the EndoWrist instruments are two inches longer; together, these two features facilitate multiquadrant access), and high-definition video technologies, high-speed networking, and image-guidance systems.

In addition a new feature, called TilePro, allows surgeons to import and view a variety of video images without leaving the console. All these new features, fully integrated into the da Vinci System, transcend the limitations of both open surgery and thoracoscopy, expanding the surgeon's capabilities and offering a minimally invasive option for many complex procedures, such as major lung resections.

A critical point of the robotic lobectomy is the placement of the trocars and the surgical cart in the right position to obtain the best performance of the operation. In video-assisted thoracic surgery (VATS) the baseball-diamond principle is generally accepted as the concept guiding trocar placement. In robotic lobectomy a different planimetry is required with the trocars positioned at a greater distance from each other than they normally are in standard thoracoscopic procedures. VATS exploration before the operation can yield important information that would markedly alter the treatment strategy. This strategy is particularly helpful when the fourth arm is used (to evaluate spatial relationship of the lobes and the location of the lesion) for the physical orientation and the optimal working angles between instruments, lesions plane, and the angle of the vision. The best positioning of the system and the robot arms is established to obtain an excellent view of the chest cavity without arm impingement and interference. Patients are prepared and draped for a posterior lateral thoracotomy so that they can be converted in the event of intraoperative complications or in case a video robot lobectomy is not feasible. In addition to robotic-thoracoscopic instruments, a full thoracotomy instrument-set is available at the operating table.

Robotic lobectomy: surgical sequences

Video robotic lobectomy follows the standard surgical steps of open thoracic surgery and implies the isolation and resection of the vascular and bronchial hilar elements. Usually the artery is dealt with before the vein and eventually the bronchus is resected; however, priorities are not strictly set. Like other thoracic procedures single-lung anesthesia is achieved by a double-lumen endotracheal tube. The patient is placed in a maximally flexed lateral decubitus position. The standard layout is as follows:

- 1st incision (camera port): 7th–8th intercostal space (ICS) at the mid-posterior axillary line.
- 2nd incision (arm port 2): 6th–7th ICS at the posterior axillary line; above the diaphragm posteriorly to the scapula tip.
- 3rd incision (arm port 1) (utility incision, 3 cm length): 4th–5th ICS anterior axillary line. The utility incision location varies depending on the lobe of interest and the hilar structures. For the upper lobectomy it is at the level of the superior vein in the midaxillary line (usually 4th ICS). For middle and lower lobectomies, the incision is placed one intercostal space lower.
- 4th incision (arm port 3): 5th–6th ICS inferior to ausculatory triangle.

When the fourth arm is not suitable an additional small incision is made (between the service entrance and the three-dimensional scope) for the assistant surgeon to insert conventional endoscopic instruments.

Once the incisions have been made, the da Vinci (surgical cart) is positioned at posterior, and the patient's head is brought into position from the posterior aspect of the patient, with the center column at an approximately 45-degree angle with respect to the longitudinal axis of the patient. A 30-degree scope angled down (generally preferred) is introduced through a 12-mm trocar and secured to the camera arm. The positioning of the instrument arms and the remaining access incisions are accomplished under direct vision.

Hilum dissection

The dissection is performed by using 1 or 2 Cadiere forceps and the permanent spatula or hook (preferably) attached to the electrocautery. Individual isolation of the hilar structures proceeds with dissection around the hilar vessels and bronchi performed through a combination of cautery and sharp and blunt dissection.

Arterial step

When the fourth arm is available, a Cadiere forcep–EndoWrist for lung retraction (through port 2), a Bi-polar fenestrated forcep–EndoWrist (through port 3), and a monopolar hook or Spatula-Endowers (through the utility incision) are used for the artery dissection. If the interlobar fissure is complete or nearly complete, the incision of the visceral pleura with electrocautery or blunt dissection with a pledget mounted on the Cadiere forceps allows the pulmonary artery to be easily identified. When the vessels are sufficiently dissected, two blunt-tipped Cadiere forceps are used to isolate the pulmonary artery. Then, a sling is passed. The dissection begins after vein isolation for upper and middle lobectomies; the arteries are isolated and often taken separately, for lower lobectomies. Thanks to the wrist instruments, the suture may be performed by using double-tie linen 2.5 or by vascular stapler introduced through the posterior access incision, when an upper or middle lobectomy is performed or through the anterior utility incision for the lower lobectomy. Innovative robotic clips are now available for pulmonary vessels (Hem-o-lok, WECK, TFX Medical, High Wycomb, United Kingdom). These clips may be delicately applied

3

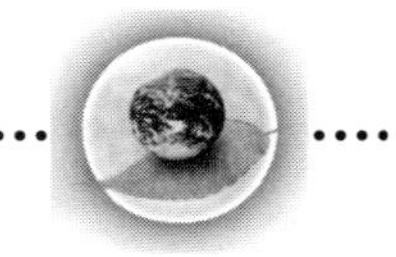

The Tropical Savanna Biome

General Overview

Savanna refers to a type of vegetation where grasses form a complete groundcover and usually there is an upper story of scattered trees, shrubs, or palms. The word *savanna* comes from a Carib word for a treeless plain covered with grass. The Spanish adopted the term when they colonized the West Indies, and they introduced it to other western languages. Today, among vegetation scientists, the meaning has changed only slightly: *savanna* refers to a type of vegetation where grasses form a complete groundcover; an upper story of scattered trees, shrubs, or palms is usual. The term is reserved mostly to describe tropical vegetation, but it can be applied to vegetation of similar structure in temperate zones. Some scientists use the term *savanna landscape* to describe a group of interrelated grassland, gallery or riparian forest, swamps, or marsh ecosystems. The popular image of tropical savanna comes from the African part of the biome, where vast grasslands (see Figure 3.1) dotted with umbrella-shaped deciduous trees and thorny shrubs (see Plate IX) are inhabited by herds of large grazing animals. Not all savannas are like this, however.

Geographic Location

The grasslands of the tropics, generally known as savannas, lie just poleward of the Tropical Rainforest and Tropical Dry Forest Biomes and form a transition between those forests and the deserts of the subtropics (see Figure 3.2). The boundary between forest and savanna is often abrupt. As distance from the equator increases, different types of savanna may form distinct zones, a pattern that results from the

Figure 3.1 General vegetation profile of an African savanna. *(Illustration by Jeff Dixon.)*

increasingly long dry seasons experienced and increasingly less total annual precipitation received the closer an area is located to the permanent high-pressure cells centered near 30° latitude in both hemispheres.

The major expression of the biome occurs in Africa, where savannas surround the rainforest on three sides (north, east, and south). South America has considerable area occupied by tropical savanna. North of the Amazon rainforest, the Llanos and other, smaller patches of savanna such as the Rupununi-Roraima savannas can be found. South of the Amazon forest, on the uplands of Brazil, is a much larger expanse known as the *cerrado*. The South American savannas look quite different from those of Africa and contain different plants and animals. A third continent with a significant tropical savanna region is Australia. As is true of most of Australia's vegetation types and animal life, its tropical savannas are inhabited by plants and mammals not found naturally anywhere else in the world.

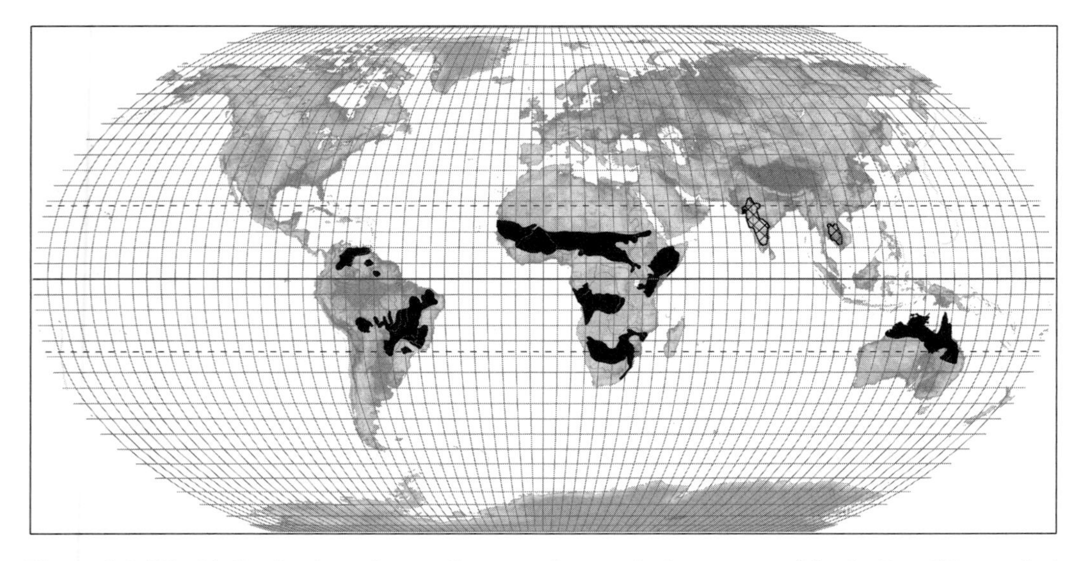

Figure 3.2 World distribution of naturally occurring tropical savannas. *(Map by Bernd Kuennecke.)*

Savannas in tropical Asia are most likely "derived" and not natural savannas. This means that they are the result of human activities such as logging, burning, and the grazing of livestock, any or all of which altered tropical dry (or deciduous) forests that once grew in the area. Since they are not considered natural, the tropical savannas of India and Southeast Asia are not described in this book.

Climate

Tropical savannas occur in areas with distinct annual rhythms based on alternating wet and dry seasons. Abundant precipitation during the rainy season makes this is a type of tropical humid climate known as the tropical wet and dry climate (Aw in the Köppen climate classification system), or often simply the tropical savanna climate. The rains come during the high-sun period, when the Intertropical Convergence Zone (ITCZ) is positioned nearby. The ITCZ is the place where the trade winds of the Northern and Southern Hemispheres meet and force the air to rise and generate rainfall. Its position shifts during the year as the vertical rays of the sun migrate between the Tropic of Cancer and the Tropic of Capricorn. The low-sun season, when the vertical rays strike in the opposite hemisphere, is the dry season. There may be three to four months when little or no precipitation falls (see Figure 3.3). Total precipitation within the biome varies from 40–60 in (1,000–1,500 mm) to as little as 20 in (500 mm) a year.

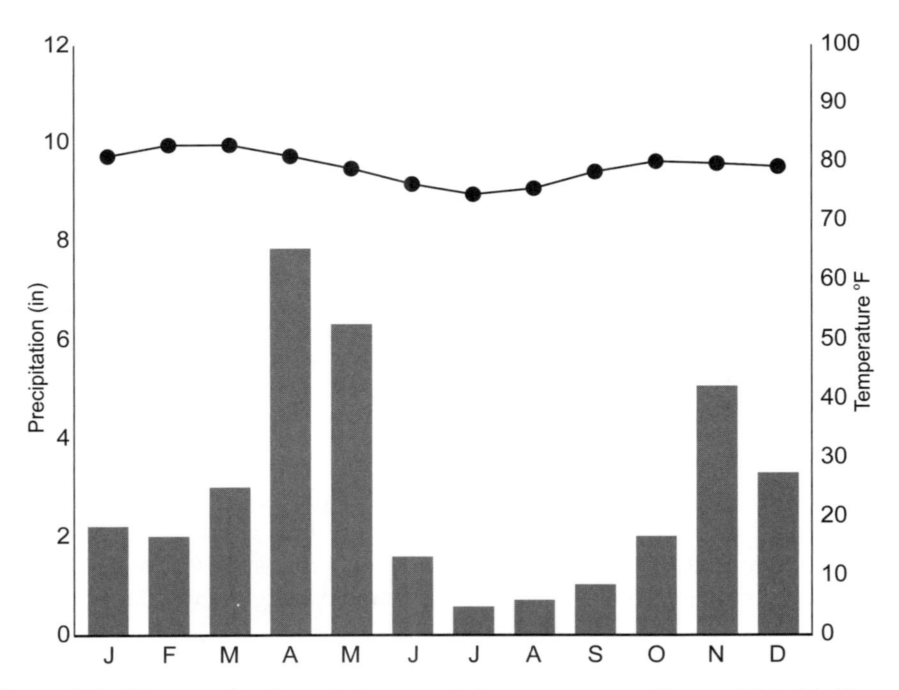

Figure 3.3 Climograph of tropical wet and dry or savanna climate: Nairobi, Kenya. *(Illustration by Jeff Dixon.)*

The tropical nature of the climate is apparent in the year-round warm temperatures, with frost being an unknown or extremely rare event. The temperature difference between the warmest month (usually at the end of the dry season) and the coolest month (during low sun) is only 5°–10° F (2.8°–5.5° C). The difference between daytime and nighttime temperatures is often greater.

Even though a close correlation exists between tropical savannas and the tropical wet and dry climate, climate does not seem to be the main reason most savannas are found where they are. Repeated burning, low nutrient soils, iron-rich hardpans close to the surface, and the impacts of large grazing and browsing mammals are also factors in the presence and maintenance of many of the world's tropical grasslands.

Other Contributing Factors

A variety of factors other than climate contribute to the formation of the tropical savannas. These factors include fire, grazing, and edaphic conditions, and are discussed below.

Fire. Fire is such a common feature of the savanna environment that its presence is sometimes included in the list of factors that make a savanna a savanna. Natural fires are part of the seasonal rhythm of life. Sparked by lightning, they usually occur just as the rainy season begins. The rain of the earliest storms may evaporate before reaching the ground, so lightning strikes last year's accumulation of dead leaves and dry stalks. The continuous cover of dried grass carries the flame across the landscape. Both plants and animals show adaptations to the annual grass fires. Trees have thick bark that, while it may get scorched, protects the living tissue beneath it from damage. Some trees and shrubs have buds below the ground surface—where they are insulated from the heat—and are able to resprout quickly. Grasses and subshrubs have underground structures that are similarly protected from all but the hottest of fires. Burning reduces plant tissues to their component minerals and deposits them as a layer of ash on the ground. The ash is a short-lived coating of fertilizer that promotes rapid new growth in the recovering plants.

Among animals, some find shelter below ground while a fire races through the savanna vegetation above them. Others flee ahead of the flames. Not all escape, however, and the dead and dying are food for a number of predators and scavengers. Some birds and mammals are actually attracted to the flames, since insects and other small animals flushed by the fire are an easy meal.

Humans have greatly increased the frequency of fires and expanded the area covered by savannas. Our earliest known ancestors originated on the savannas of Africa. Early humans may have increased the habitat best suited to them by burning grasslands at the edge of forests and gradually converting dry forests and rainforests to savanna. Some suggest that we continue to create savannas well outside the tropics, but now we call them lawns and parks—that we have some kind of inherited memory that makes us prefer grasslands with scattered trees.

Grazing. On the savannas of eastern and southern Africa, the impact of the large numbers and great diversity of hoofed animals (ungulates) appears to play a major role in producing and maintaining the habitat mosaic of the savanna landscape. In their absence, the natural vegetation would likely be a mopane woodland or miombo forest too shady for the growth of a continuous layer of grasses. Elephants are particularly destructive of trees. During the hottest part of the day they seek the shade under the woodland's closed canopy; but whether from boredom, an itch, or playfulness, they rub against the trees and can push them over. Seeking food, elephants will tear off whole branches with their trunks (see Figure 3.4). Their tusks rip apart the bark, which can kill trees. So even a small herd of elephants can open the woodland canopy and let sun reach the ground. This lets tall, sun-loving C_4 grasses invade and turns the woodland into savanna. Once there is a continuous groundcover of grass, fire easily spreads through the area. Fire kills tree saplings and seedlings and thereby prevents the regeneration of many woody plants. New shoots of grasses attract grazers, whose nibbling and trampling of tree seedlings further inhibits the regrowth of woody plants. Browsing animals consume the foliage, fruits, and twigs of any trees or shrubs tolerant of fire and keep stands of woody plants in check.

The vegetation change is reversible, however. Heavy grazing pressure or overgrazing can break up the continuous grass cover and result in patches of bare

Figure 3.4 Elephant carrying branch torn from savanna tree. *(Photo by author.)*

ground. Under these conditions, fire is less able to move across the land; and shrubs and trees invade the open places. The invading species of woody plants are usually armed with thorns to help fend off browsing animals. Each herbivore has its own preferred habitat and forage type within the savanna landscape. To conserve as many species as possible in the remnants of the great African savannas that survive today, wildlife and park managers must allow for the interplay among plants, animals, and fire. South American and Australian savannas lack the ungulates so characteristic of Africa's grasslands. Small mammals and invertebrates are significant plant eaters, but they do not seem to affect the savanna landscape in the major way African ungulates do.

Edaphic conditions. Edaphic factors are those properties of the soil or parent material that affect the growth of certain types of plants. Negative impacts arise from poor drainage that leads to waterlogged conditions, including the presence of a hardpan (laterite) that prevents root growth; excessive drainage in sandy soils that keeps the ground too dry for most trees; low nutrient content that is a consequence of either leaching or original bedrock type; and soils that are toxic because of a high concentration of aluminum, such as found in parts of the Brazilian Highlands. Trees, especially, are sensitive to edaphic conditions, and the scarcity of trees in some savannas is due to their inability to tolerate such unfavorable environments. Grasses with their shallower root systems, seasonal diedown, and lower demand for nutrients flourish in the sunlit habitats available when trees are sparse.

Some woody plants adapt to the very conditions that eliminate most trees and shrubs. Certain palms, for example, can withstand having their roots in water-saturated soils. Some palm savannas exist because of such edaphic conditions (see Figure 3.5), although other types of palm savanna develop where fire is a recurrent feature of the environment. In Brazil, a number of woody plants have evolved means of protecting vital tissues from aluminum-based compounds, and savanna woodlands exist on soils that would be toxic to most trees and shrubs.

Vegetation

The typical grasses of the Tropical Savanna Biome are bunchgrasses some 2–4 ft (50–120 cm) high. The blades are tough and contain many silica bodies. Savanna grasses use the C_4 photosynthetic pathway (see Chapter 1). C_4 plants are most efficient at assimilating the energy of the sun at high temperatures. They have a high degree of resistance to drought, since C_4 plants take in carbon dioxide more rapidly than do C_3 plants and therefore do not have to leave their stomata open as long. Water is lost from a plant (transpired) through the stomata, so the closing of these pores acts to conserve plant moisture.

In addition to grasses, the herb layer of tropical savannas is often rich in sedges and forbs. Many of the forbs are members of the pea or legume family. Since

Figure 3.5 Palm savanna formed in water-logged conditions, Rio Grande do Norte, Brazil. *(Photo by author.)*

legumes have symbiotic relationships with *Rhizobium* bacteria that fix nitrogen from the atmosphere, they are able to thrive in the often nutrient-poor soils of the tropics.

Trees of the savannas, with the exception of the huge baobabs of Africa and Australia, are rarely taller than 40 ft (12 m). Most are only 6–20 ft (2–6 m) tall. They have entire leaves that often have thick waxy coatings (cuticles) and sunken stomata to lessen water loss. Some have leaves that are thick and leathery in texture. Most trees and shrubs are deciduous, but some drop one year's leaves at the same time that new buds are opening, so they may appear to be evergreen.

A thick bark often defends savanna trees against fire. Many can resprout from their roots, another survival tactic in face of frequent burning. In Africa, where grazing by large mammals is a major factor, many woody plants are armed with thorns. These same trees and shrubs are apt to have tiny leaves and very deep taproots.

Other growthforms commonly found in tropical savannas include subshrubs and annual forbs and grasses, all well adapted to strongly seasonal environments. The subshrub dies down at the end of one rainy season and produces new woody stems at the beginning of the next. A woody underground structure (xylopodium) stores energy and nutrients to allow for rapid regrowth (see Figure 3.6). Annuals avoid dealing with drought by existing as dormant seeds during the

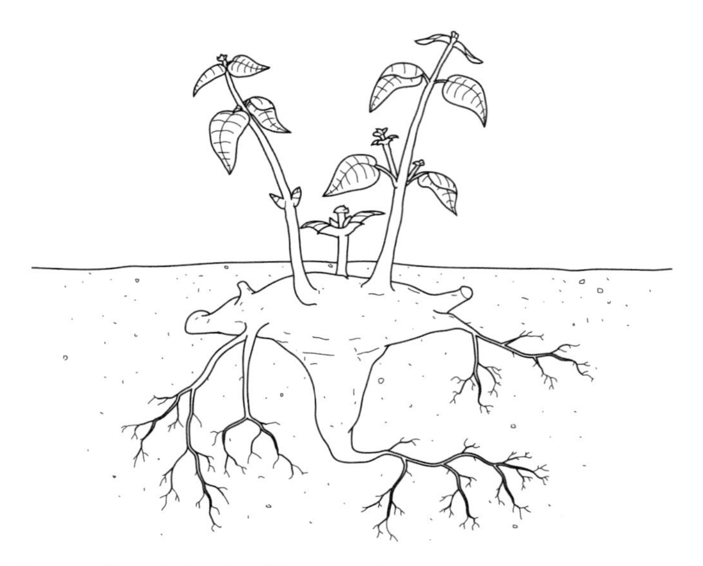

Figure 3.6 Form of a subshrub with an underground storage organ or xylopodium. *(Illustration by Jeff Dixon. Adapted from Monasterio 1983.)*

nongrowing season. The seeds will germinate quickly with the onset of rains, and the plants will complete their life cycles before the next dry season sets in. (See the appendix to this chapter for lists of plants characteristic of specific parts of the Tropical Savanna Biome.)

Two flowering seasons are usual in the Tropical Savanna Biome. Some plants bloom at the very beginning of the rainy season even before their leaves have fully developed. Other species bloom in the middle of the rainy season or toward its end.

Scientists recognize different types of savanna according to the spacing and frequency of trees or other woody plants. There is no commonly agreed-on classification for savannas, but Table 3.1 describes some of the more frequently used types. The common factor in all is a continuous cover of grasses.

Table 3.1 General Types of Tropical Savanna

Savanna Type	Description
Savanna Woodland	Many trees with open canopy so that light reaches ground
Park Savanna	Trees in clumps with expanses of grass between clumps
Tree Savanna	Trees scattered apparently randomly across the area
Shrub Savanna	Shrubs are the dominant woody plants on the grasslands
Palm Savanna	Palm trees, scattered or clumped, are the dominant woody plants
Grass Savanna	Trees generally absent, except in gallery forests

Soils

Soil formation. Soils in humid tropical climates are exposed to high rates of leaching, the dissolving and washing down into the soil column of any soluble compound. The typical soil-forming process is one of laterization (see Figure 3.7). The amount of leaching that has occurred is related to the age of the land surface. The tropical savannas of Brazil and western Africa lie on very ancient surfaces—exposed continental shields—that for millions of years have been affected by high temperatures and relatively high amounts of rainfall. The bedrock has been deeply weathered, and most soluble compounds, including even silica compounds, have been removed. Since plant matter decays rapidly in constant warm, moist conditions, essentially no humus is available to help bind soluble bases (which are key plant nutrients) even temporarily. Left behind is a soft residue high in iron oxides and aluminum oxides and lacking in most plant nutrients. The high amount of iron in the soils gives them a bright red or yellow color (see Plate X). Concentrations of aluminum can be high enough that the soil is mined as an ore (the ore of aluminum metal is called bauxite). In parts of the Brazilian Highlands, aluminum

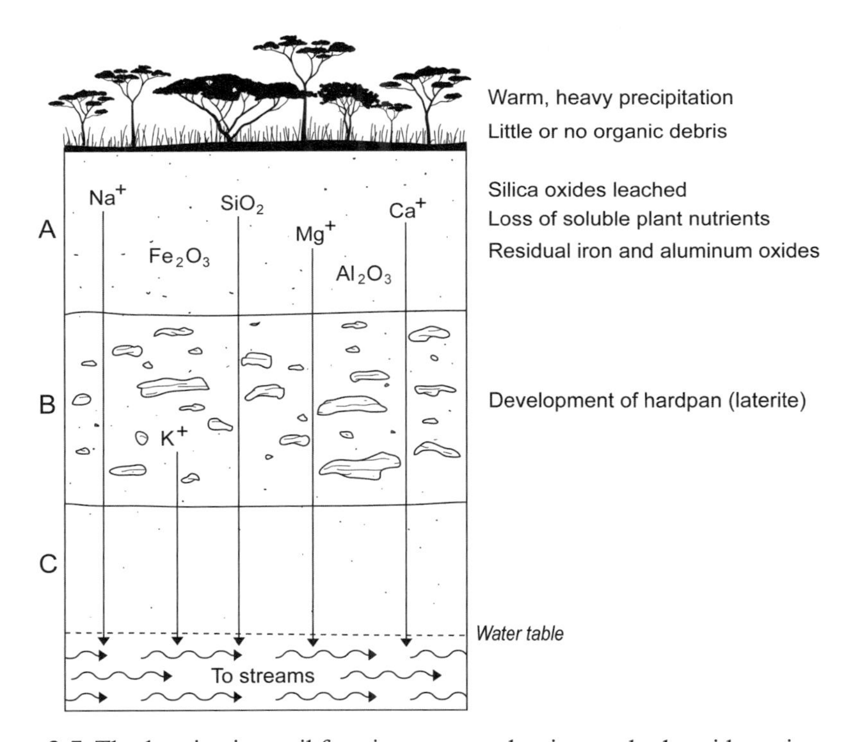

Figure 3.7 The laterization soil-forming process dominates the humid tropics on ancient rock surfaces. Soluble nutrients and silica oxides are leached from the soil column leaving only oxides of iron and aluminum and resulting in a naturally infertile soil. *(Illustration by Jeff Dixon.)*

concentrations are so high that they are toxic to many plants. Native trees and shrubs have special adaptations to withstand these challenging conditions.

Soils developed on ancient surfaces composed of granitic rocks tend to be quite acidic (average pH is 4.9), further reducing the ability of the soil to hold nutrient bases. Another problem for plant life is the repeated wetting and drying of the iron-rich soils that can turn them into a brick-hard layer known as laterite. This hardpan prevents root penetration and halts the percolation of water during the rainy season, creating waterlogged soils or even temporary pools of standing water. The existence of laterite can be one reason for the lack of trees in some tropical savanna areas.

In some parts of the tropics, weathering and leaching have been neither so prolonged nor extreme. In well-drained areas and regions where the crystalline bedrock is not as old or where it has been more recently exposed than described above, soil-forming processes more like those of temperate regions (podzolization) are able to take place. Over much of eastern and southern Africa, for example, the savanna soils are less acidic (pH is 6.2) and somewhat more fertile than on the more ancient shield surfaces.

On base-rich bedrock, such as basalt or limestone, soils develop that are high in essential plant nutrients such as calcium and magnesium (both are bases). They typically have a high clay content that makes them swell during the wet season and shrink and crack during the dry season. Such soils have relatively high natural fertility and tend to be associated with grass savannas.

Soil types. The typical, iron- and aluminum-rich, low-nutrient soils of the ancient plateaus of Africa and South America are classified in the U.S. Soil Taxonomy as oxisols. "Oxi" refers to the oxides of those two elements that are so concentrated by tropical soil-forming processes. Similar soils are also found in the Tropical Rainforest Biome. In other classifications, they are known as latosols or lateritic soils.

Where drainage is better or younger, crystalline rock forms the parent material of the soils. According to the U.S. soil system, these soils are ultisols. "Ulti" means ultimate or last: the last of the nutrients have been removed by leaching.

The more fertile soils developed on basalt or limestone and still containing abundant plant nutrients are among the vertisols of the U.S. Soil Taxonomy. "Vert" means turn. The swelling and shrinking of the clay mixes or turns the soil so that no true horizons can form. In other soil classification systems, they are more vividly named "black cracking soils."

Animals

There can be no single description for animal life in the world's various tropical savannas. The popular view of savanna life comes from the game parks of eastern and southern Africa, where herds of large hoofed mammals—zebra, wildebeest, antelopes in delightful variety, as well as the great elephants, rhinoceroses, and giraffes—dominate the scene. They truly represent one of the spectacles of nature, but they are not typical of savannas anywhere else. In most savannas, small mammals—rodents, rabbits, and hares—are more characteristic. In Australia, small marsupials resemble rodents and hares.

Small mammals are usually nocturnal and seldom seen. Many seek refuge below ground during the warmest parts of the day: they are fossorial. Burrows also provide protection from the grass fires that are frequent at the end of the dry season for those that cannot run away from the flames. The small mammals tend to be omnivores, consuming green leaves, fruits, seeds, roots, and tubers, as well as insects and other invertebrates. A considerable number focus almost exclusively on invertebrates, such as the extremely abundant termites and ants. Larger predators consume reptiles, birds, and rodents. Only Africa has a large number of both feline and canine carnivores that regularly hunt large mammals.

Scavengers play important roles in recycling animal matter. Vultures are conspicuous components of the scavenger community in tropical savannas.

Mammals of the tropical savannas have evolved ways to deal with intense sunlight, high air temperatures, and limited amounts of water. Large mammals may not sweat, but instead pant to lower their body temperatures. Panting lets them cool down without losing large amounts of water. Others reduce water loss from

Ant and Termite Eaters: Examples of Convergent Evolution

Among the strangest looking mammals of the Tropical Savanna Biome are those adapted to eating termites and ants. Both Africa and South America have such animals, but they are not even distant relatives. Scientists classify them in entirely different orders, because they are no more closely related than dogs and squirrels or bats and cows. Africa has pangolins, also known as scaly anteaters (order Philodota), and aardvarks (order Tubulindentata). South America has giant anteaters (order Xenarthra) (see Figure 3.8). One of the interesting things about these animals is that their body and head shapes, very long tongues, and the large claws on their front feet make them look like each other. Since they eat the same foods and live in similar environments, they have evolved similar adaptations—that is, they have converged toward each other—through natural selection.

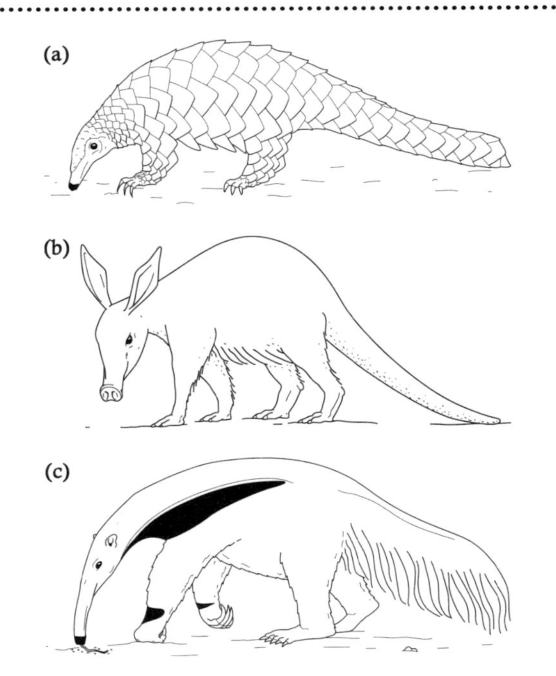

Figure 3.8 Similarity of form among termite-eating animals: (a) pangolin or scaly anteater of Africa, (b) aardvark or antbear of Africa, and (c) giant anteater of South America. *(Illustration by Jeff Dixon.)*

their bodies by having the physiological means to concentrate their urine or reabsorb water in the large intestines so they eliminate only very dry feces. African antelopes such as the dikdik, impala, and eland possess such adaptations.

Some savanna mammals reduce water loss and keep cool through behavioral adaptations. During the hottest times of day, they may retreat underground to avoid high temperatures. Simply standing in the shade of a tree is another strategy. Some of the large African animals must remain out in the sun, but they try to present as small a part of their body as possible to the strongest rays. Zebras, for example, stand with their rumps facing the sun. Other mammals depend on the light coat color common among savanna animals to reflect sunlight and thereby reduce the amount of solar energy absorbed. Feeding at night allows animals not only to be active when temperatures are lower and humidity higher, but also to forage when the moisture content of leaves is highest. During the dry season, some of the more mobile animals become nomadic and wander in constant search of areas receiving some rain, while other species migrate along pathways used for generations to distant sites with reliable waterholes.

On all three continents with natural savannas, mammals share the land with large, flightless birds known as ratites. The Ostrich inhabits African savannas (see Plate XI), the Emu occurs in Australia, and rheas live in the South American grasslands. Other birds, possessing the ability to fly, are both abundant and diverse throughout the tropical savannas. An impressive number of larger predators and scavengers such as hawks and vultures are typical, as are smaller songbirds, many of which are seed-eaters. Both South America and Australia boast large numbers of fruit-eating parrots.

Reptiles are also abundant and diverse in tropical savannas. Vipers, pit vipers, and boas; geckos and skinks; and tortoises are characteristic in all savannas. Chameleons occur in Africa, iguanas in South America.

Frogs, tree frogs, and toads live in areas where open water is available all year or at least during every rainy season. Like the small mammals, most amphibians are active only at night, and some only during the wet season. During the dry season, they remain in moist underground chambers. A few species spend their entire lives below ground.

Probably the most important animal in the savanna ecosystem is a tiny invertebrate that is largely unseen, but occurs by the millions: the termite. Although termites are visible only during the short period when winged individuals fly forth to form new colonies, the structures some species build above ground—their mounds or termitaria (see Figure 3.9)—are conspicuous elements of all savannas. The towers act like ventilation shafts to exchange oxygen and carbon dioxide between the termite nests below and the atmosphere outside. The hardened clay construction material also serves as a first line of defense for the colony from predators.

Breaks are quickly filled by soldiers that employ a type of chemical warfare by injecting toxic saliva with their bites. Termite specialists such as anteaters and aardvarks need strong claws to dig through mounds and long tongues to extract the tasty fat- and protein-rich morsels inside.

Figure 3.9 Termitaria or mound of *Macrotermes* termites in South Africa. *(Photo by author.)*

Termite Farmers

Termites feed on living, dead, and decaying plant matter, including that found in dung. They cannot digest cellulose but depend on microbes in their gut or fungi living in their termitarias to process the material into forms of energy and nutrients they can utilize. One group of termites in Africa, the *Macrotermes*, build huge mounds riddled with tunnels and laced with comb-like surfaces on which fungus grow. The towering mound provides nearly constant temperatures and humidity to aid the growth of the fungi. The termites bring dead plant material—which the termites themselves cannot digest—to feed the fungi and then the termites consume the fungi. The fungus combs are like farms and the termites are the farmers. In South America, leaf-cutting ants do the same thing.

Termites can be compared with the prairie dogs of the prairies of North America (see Chapter 2). As keystone species, termites create unique microhabitats, affect nutrient cycles and soil structure, and provide food and shelter for a number of other species.

Construction of the mounds and towers requires bringing soil up from depth and the procedure recycles nutrients from the subsoil. The mounds become areas of fine-grained, nutrient-rich (particularly high in calcium and nitrogen) soil. Local human farmers seek out abandoned mounds as sites for growing crops. Certain wild shrubs also prefer to grow on mounds so that a group of plants distinct from the surrounding grassland can develop on them.

Major Regional Expressions of the Tropical Savanna Biome

African Savannas

On a world map, the savannas of Africa appear as a continuous belt circling the Congo rainforests, and, indeed, savanna vegetation covers 60 percent of sub-Saharan Africa. However, a closer look reveals at least three separate savanna regions (see Figure 3.10). The West African savannas form a thin strip north of the equator (5°–15° N) from the Atlantic coast of Senegal and southernmost Mauritania east to the Red Sea coast of Sudan. The East African savannas straddle the equator from Somalia (16° N) south through eastern Ethiopia and Kenya into Tanzania (9° S). And the Southern African savannas occupy an area in the Southern Hemisphere from Namibia to Mozambique and south into the Eastern Cape Province of South Africa (from 18° S in Namibia to as far as 34° S in South Africa). Of these three regions, the eastern and southern savannas are most similar in plant and animal life. The West African savannas, geographically isolated from the other two African savannas, are somewhat different.

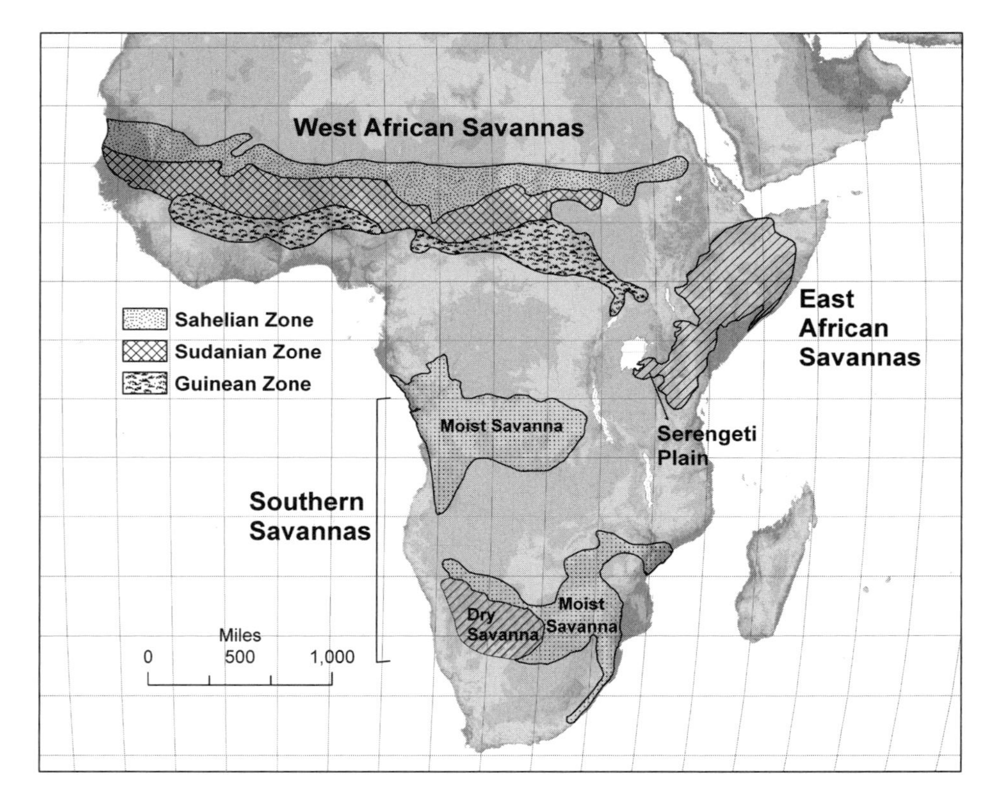

Figure 3.10 The three major regions of tropical savanna in Africa. The latitudinal zonation of the West African savannas is shown. *(Map by Bernd Kuennecke.)*

West African savannas. The West Africa savanna region is divided into vegetation belts aligned according to latitude. This zonation reflects the shorter wet seasons and overall lowered amounts of rainfall experienced with greater distance from the equator. West African savannas lie directly south of the Sahara and form a clear transition from the arid conditions of the Sahara to the year-round precipitation conditions of the Congo rainforest (see Figure 3.11).

The northernmost and driest zone is known as the Sahel, an Arabic word meaning "shore." The Sahel is the shore of that great sea of desert, the Sahara. This region receives less than 20 in (500 mm) of rain between May and September in those years when the ITCZ migrates that far north. The dry season lasts six to eight months, during which time hot, dry winds known as the harmattan are apt to blow sand and dust in from the Sahara. The Sahel is vulnerable to yearlong drought during El Niño years and other global cycles of climate change. Further contributing to aridity in the region are the highly permeable soils through which rainwater rapidly drains. The soils were developed on sedimentary deposits and former (Pleistocene) lake beds. Very few sources of permanent surface water exist.

The vegetation of the Sahel is that of a dry grassland with a sparse upper layer of scattered bushes and small trees, generally less than 15 ft (5 m) tall. Most of the grasses are annuals, which are better able than perennial species to withstand long dry periods. The bushes are primarily different kinds of thorny acacia. Most plants have no common names in English. The vegetation has been greatly altered by both human activities (grazing of cattle and goats, dry-farming techniques for millets and sorghums, and firewood collection) and long-term climate changes, such as during the period of desertification that caused widespread starvation of livestock and people in the early 1970s.

The larger native mammals—such as scimitar-horned oryx, bubal hartebeest, cheetah, and lion, that once roamed the Sahel have essentially disappeared as a

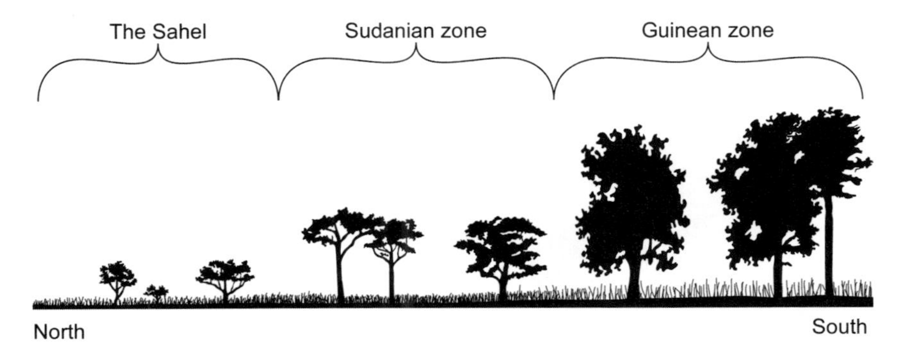

Figure 3.11 North-south transect across the West Africa savannas showing changes in vegetation profile that reflect changes in total annual precipitation. *(Illustration by Jeff Dixon.)*

result of overhunting. The introduction of rifles and four-wheel drive vehicles spelled the end for many of the larger animals. Among smaller mammals are several rodents that are found only in this part of the biome, including several gerbils and a zebra mouse.

South of the Sahel is the Sudanian zone, an area where precipitation averages between 20 and 60 in (600–1,000 mm) a year. The Sudanian zone crosses Africa from Nigeria and Cameroon through Chad and the Central African Republic into Sudan, interrupted only by the vast Sudd swamps in the upper reaches of the Nile. The rainy season runs from April to October. Soils tend to be ultisols and alfisols, with oxisols appearing only in the east.

The Sudanian zone consists of dry savanna woodland with a greater diversity of tree species—and somewhat taller woody plants—than the Sahel. Dominant are deciduous *Terminalia*, *Combretum*, and *Isoberlinia* trees and shrubs, although acacias are common. All of these small-leaved, deciduous trees generally reach 15–20 ft (5–7 m) in height. Towering above them to heights of 50 ft (15 m) may be enormous baobabs, with their swollen trunks and odd, upside-down appearance (see Figure 3.12). So-called elephant grasses may grow 10–12 ft (3–4 m) tall. The threats to the vegetation are the largely the same as in the Sahel: overgrazing, burning, farming, and wood collection.

Figure 3.12 A baobab towers above the typical trees of the savanna in Kruger National Park, South Africa. *(Photo by author.)*

Baobabs and People

In the West African savannas, baobabs grow where people live, so much so that it appears the giant trees were introduced to the drylands north of the African rainforest by humans centuries ago. Baobab is an important fruit-bearing tree in rural settlements, and the shade cast by large specimens makes them excellent "palaver trees," sites where the community assembles to debate and make decisions on local issues. The mealy pulp of the fruit is commonly added to the morning porridge. The pulp is soaked in hot water to remove the seeds, which are discarded. This practice does not destroy the seeds, but actually promotes germination. Refuse areas around a village become seed beds for baobab, and few grow well elsewhere. In addition to the better soils and shallower water tables that shifting agriculturalists select for occupation, the villages offer a dry, fire-free habitat. Young baobabs are destroyed by fire. Large specimens, however, tolerate fire and, in their adopted land, tend to occur in the open fire-swept spaces of fallowed fields and abandoned villages where they cannot reproduce.

A large baobab might seem like an excellent location to establish a new village, because it offers fruit and shade. However, ownership and use practices may prevent this. People gain the right to use (usufruct) certain valuable trees, a right that can be passed to descendents irrespective of the ownership or use of the land. To avoid possible conflict, new settlements stay away from "owned" trees and encourage the growth of new saplings whose fruits they will be able to claim as their own.

Though largely dependent on people for their dispersal and protection, baobabs do grow beyond human habitation. Baboons and chimpanzees like the fruits, too, and carry them away from the villages, depositing the seeds in their droppings. On dry cliffs where wildfires are rare, the seeds will germinate far from human occupation sites.

Small remnant herds of elephant survive in a few protected locations, as do West African subspecies of buffalo, giraffe, and giant eland. Roan antelope are similarly rare. Black and northern white rhinoceroses have long been lost from the area. Large predators probably persist in larger numbers than in the Sahel zone.

In addition to habitat loss from direct human encroachment, potential threats to animals arise from efforts to harness the natural waterways for human use. Drainage of wetlands, damming of rivers, and diversion of streams will alter the natural flooding patterns of vital water sources.

The southernmost part of the West African savanna belt is the Guinean zone. This is a moist savanna woodland, where annual rainfall averages 40–60 in (1,000–1,500 mm). The rainy season lasts from April to October. Tall grasses dominate, but the landscape contains many patches of *Isoberlinia* and other types of woodland in which trees reach heights of 40–50 ft (12–25 m). Gallery forests along streams are conspicuous and important habitat for the animals of the region. The southern margins of the zone are not natural savanna but have formed as result of repeated burning of semideciduous tropical forests by herding and farming peoples. The conversion of forest to savanna proceeds at a rapid rate.

The Guinea savannas are home to many of the same large mammals as the two other zones in the West African savanna. The presence of gallery forests, wetlands, and patches of tropical forest, however, increases the diversity.

East African savannas. Savanna occurs from 16° N to 9° S latitude and covers much of the Central African Plateau east of the Great Rift Valley and Ethiopian Highlands from eastern Ethiopia and Somalia through Kenya and into Tanzania. This part of the biome coincides with the World Wildlife Fund's Somali and Northern *Acacia–Commiphora* bushlands ecoregions and includes the famous Serengeti Plains and Ngorongoro Crater. Conditions are driest in the north, where the East African savannas converge with desert shrublands. The climate is wettest in the south, where the savannas give way to miombo and other tropical dry forests.

Most of the region experiences two rainy seasons and two dry seasons a year because of its position on or close to the equator. In its yearly passage, the ITCZ crosses a given area once as it migrates toward a tropic and then some time later, when it returns to the equator. The rainy seasons are not of equal length, and local peoples have long distinguished between the long rainy season (March to May) and the short rainy season (mid-October to December). At neither time, however, are the rains completely reliable; and periods of rain may be brief or fail altogether. Generally speaking, total annual precipitation averages about 20 in (500 mm).

With the notable exception of the Serengeti, the tropical grasslands of East Africa are tree and bush savannas composed of shrubs and small trees that lose their leaves during the dry season. Most have thorns. Shrubs may be 10–15 ft (3–5 m) tall, and scattered trees—often *Acacia tortilis* and *Terminalia spinosa*—stand above them, attaining heights of about 30 ft (9 m). Acacias of one type or another occur nearly everywhere. Dense thickets of corkwoods and grewia bushes are also characteristic. Giant baobabs, native to these savannas, may tower above all other woody species. The grasses tend to be annuals and short-lived perennials.

The dry part of the savanna biome that extends north into the Horn of Africa was once part of a great desert complex that connected it to the Namib and Kalahari in southwest Africa and to the more northerly desert belt sweeping across the Sahara and through the deserts of southwest and central Asia. As a result, many desert-adapted species are found in the Somalian and northeast Kenyan savannas. Now isolated from the great desert belt, this ancient, stable area is a center of endemism for plants, reptiles, and mammals. Among mammals found nowhere else are several antelopes (see Table 3.2), some endemic gerbils, and an endemic subspecies of warthog. The range of the rare Grevy's zebra, better able to tolerate

Table 3.2 Antelopes Endemic to the Dry Savannas of Eastern Ethiopia, Somalia, and Northeast Kenya

Common Name	Latin Name
Dibatag or camel antelope	*Ammodorcas clarkei*
Beira	*Dorcatragus megalotis*
Hirola	*Damaliscus hunteri*
Speke's gazelle	*Gazella spekei*

drought than its more common relative the plains zebra, stretches south into the region. Small remnant populations of the endangered African wild ass (*Equus africanus somaliensis*) may also survive here.

Kenya and surrounding areas of southeast Sudan and northeast Uganda lie in a mixing zone for the desert-adapted species of the northern East African savannas and the more tropical species that dominate in the moister, southern parts of the region. This is the classic or "typical" acacia savanna of the western imagination,

The Upside-down Tree of Life

One of the more conspicuous and fascinating elements of African savannas is the great baobab tree with its massive trunk and array of stubby branches (see Figure 3.12). Drawing water into its tissues during the rainy season, the largest baobab trunks can swell to almost 50 ft (15 m) in diameter. The branches—leafless during the dry season—looked like roots to some travelers, who came to call the baobab the upside-down tree. But there are even more colorful nicknames for the tree.

The hairy, egg-shaped fruit dangles from a long stem or tail, so some people call the baobab the "dead-rat tree." When the fruit falls to the ground, it is eaten by elephants, antelopes, baboons, and monkeys; and it becomes a "monkey-bread tree." The white pulp that surrounds the fruit can be made into cream of tartar, a cooking ingredient used to fluff beaten egg-whites—hence, it is sometimes called the "cream-of-tartar tree."

The pulp of the fruit and the leaves are rich in vitamin C and used medicinally by local peoples. Leaves can be eaten like spinach; seeds are edible or can be brewed into a drink like coffee. The bark and wood of the tree, though not good for construction, are fibrous and can be made into cordage for nets, basket, and snares. In times of drought, the bark and the roots can be chewed to release the water stored in them.

Galagos (bushbabies), squirrels and other rodents, lizards, snakes, tree frogs, and honeybees all find refuge in hollows and crevices in these great trees. Cavity nesters such as hornbills and parrots raise their young in holes in the baobab. Older trees, while still alive, may become hollow, creating large chambers that people have used as homes, storehouses, chapels, bars, prisons, tombs, and even an outhouse complete with flush toilet.

Baobabs can live for 2,000–3,000 years. When they die, however, they seem to disappear almost over night. They collapse into a soggy mess of decaying mush.

and fortunately it is fairly well protected in national parks and reserves. Much of the early ecological research on African savannas was conducted in Tsavo, Amboseli, and other National Parks, so much of our knowledge of savannas is based upon this region.

This part of the East African savanna is formed on the ancient bedrock of the Central African Plateau. Elevations rise from 650 ft (200 m) in the north to (3,000 ft) 1,000 m in the south and southwest. Rainfall averages from a low of 8 in (200 mm) near Lake Turkana to about 24 in (600 mm) among the coast of Kenya, but it is variable from year to year and often one or both rainy seasons fail.

The landscape is a mosaic of woodland and savanna perpetuated by elephants. It supports a legendary diversity of grazing and browsing ungulates as well as the cats, dogs, and hyenas that prey on them (see Plate XII). Scientists have long wondered how so many different kinds of mammals can live together in the same place, and some of the early ecological studies offered possible explanations. Each herbivore has its preferred habitat, food plants, active time of day, and time of year for occupying a particular location. This partitioning of the habitat and its resources by use patterns is particularly true of browsers, which consume not only different kinds of plants, but may eat from different heights depending on the size of the animal. The grazers interact in what has been called a grazing succession. Zebras seem to be the first to enter an ungrazed area. They are able to utilize low-nutrient, coarse forage, so they remove the mature stems of tall grasses and expose the shorter grasses preferred by wildebeest. Smaller antelopes, such as Thomson's gazelle, eat less and so require more nutritious plant matter. They move in just as the grasses are producing new, protein-rich shoots in response to having been cropped by wildebeest and other large grazers.

Scavengers are vital in recycling the hundreds of thousands of carcasses that result not only from predator kills, but also from old age, broken bones, and disease. Perhaps the most visible scavengers are the several kinds of long-necked, bald-headed vultures that congregate at kills and rotting carcasses (see Plate XIII). Like the many grazing and browsing mammals, each has its own role. A group of White-winged Vultures is often first to arrive. Their flashy black and white feathers may alert high flying Rüppell's Griffons to the kill, a flock of which will dismember the corpse as they rip meat from deep inside. The largest African vulture, the Lappet-faced Vulture, is often the last to arrive; but when it does, it takes over the carcass. Because it is the only one that can tear into the thickest hides, however, the others depend on its coming and wait until it has finished before they move in and clean up the leftovers.

Other birds of note include the weaver birds (family Ploceidae), which are sparrow-like birds that often build huge colonial nests. Sparrow-weavers build bottle-shape nests woven of grasses and twigs and may hang hundreds of them together in a single acacia. Individual nests are entered from below through vertical tubes. Another, the Sociable Weaver, constructs what looks like a drooping thatched roof of coarse, thorny sticks in the forks of branches in acacias or baobabs. Beneath the roof, hundreds of nests will be built (see Figure 3.13). The Red-billed Queleas,

Figure 3.13 Colonial nests of the Sociable Weaver in Kgalagadi Transfrontier Park, South Africa. *(Photo by author.)*

another weaver, is spectacular not only for its huge nesting colonies (up to 10 million nests have been described in one colony), but also because this nomadic species will suddenly descended on an area like locusts in flocks numbering hundreds of thousands, if not millions, of noisy little birds.

Several large, ground-dwelling birds are conspicuous animals on African savannas. The tallest is the Ostrich. Another is the Kori Bustard, which stands about 4 ft (120 cm) tall and is the world's largest flying bird (see Plate XIV). Kori Bustards are omnivores, consuming insects, small mammals, lizards, and snakes as well as seeds and berries. The Secretary Bird, another tall hunter, is related to hawks and is a carnivore (see Figure 3.14). About 3 ft (100 cm) from head to tail, it sports long feathers on its head that resemble the quill pens secretaries of the 1800s might have stuck in their hair. With its long legs, it scares up snakes and lizards from the grasses and stomps them to death, sometimes flinging them into the air a few times to stun them in what looks like an elaborate dance. The Secretary Bird is one of the predators attracted to grass and brush fires, where it seizes injured and fleeing animals. A strong walker, it is also a good flier.

Termites are always important members of savanna ecosystems. *Macrotermes* termites in East Africa construct tall ventilation shafts for their underground nests. The mounds support a distinct vegetation that is preferred forage for zebra, impala,

Figure 3.14 Secretary Bird, skilled predator of snakes. *(Photo by author.)*

and others. Towering as high as 20 ft (6 m) and having diameters up to 100 ft (30 m), the mounds serve as lookout posts for many animals and may become homes for dwarf mongooses, monitor lizards, and several kinds of birds. And, of course, the termites themselves are food for a number of vertebrates. Aardvarks and pangolins, honey badgers, and aardwolves (a kind of hyena) break into nests to dine on termites. When the winged forms swarm, they do so by the millions. Most are immediately consumed by waiting birds and baboons and warthogs. On occasion, humans, too, partake of the nutritious snack.

A unique and world-famous part of the East African savannas is the grass savanna of the Serengeti Plains. The soils, vertisols, derived from volcanic ash that spewed from nearby volcanoes, including the now-dormant Ngorongoro caldera and the recently active (1966) Mount Lengai, are fertile, but rainwater rapidly drains through them and it is difficult for trees to become established. Grasses—short, medium, and tall—thrive. The gently rolling terrain is dotted with rocky areas, known as kopjes, where ancient basement rock pokes through layers of ash and soil. The Serengeti is most famous as the site of an annual migration of hundreds of thousands of large mammals (see Figure 3.15). An estimated 1.3 million blue wildebeest, 200,000 plains zebra, and 400,000 Thomson's gazelles take part in the migration. They are among 23 species of hoofed mammals that fatten on the relatively nutritious grasses growing on these plains. Migratory herds arrive at the

Figure 3.15 Herds of wildebeest and plains zebra congregate during the annual migrations. *(Photo © J. Norman Reid/Shutterstock.)*

beginning of the rainy season to occupy the short-grass vegetation in the southern, driest part of the plains.

As the rainy season proceeds and dry season begins, the animals move northward into medium- and tall-grass areas. When they have grazed the grasses to stubble, they move still farther north into savanna woodlands, the wettest part of the greater Serengeti ecosystem, where they spend the dry season. The trek back to the southern Serengeti starts as the rainy season begins.

In addition to those that follow a great migration circle, some herbivores such as eland, gemsbok, and hartebeest remain in the Serengeti all year, but they are nomadic. They rotate their pastures during the rainy season, allowing time for recovery of the grasses before they return to the same place to graze again. During the dry season, their movements seem less routine, and they seek and find areas in the medium- and tall-grass areas wherever rain showers have occurred.

The predatory cats and canids of the Serengeti, as elsewhere, are territorial and remain in their home ranges all year. The rainy season with its vast herds of grazing animals is a time of plenty. The dry season, however, when the zebra, wildebeest, and others have left, can be a time of scarcity and starvation.

Outside the Serengeti, in northern and central Tanzania, the savanna occurs at higher elevations, generally 3,000–4,000 ft (900–1200 m) above sea level. This region is better watered than areas to the north and receives, on average, 25–30 in (600–800 mm) of rainfall a year. A double peak in precipitation is still typical, with the long rainy season occurring from March to May and the short rains from November to December. Some years, the two rainy seasons merge into one. From August to October, trees are without leaves. Fire and elephants turn woodland into grassland. Soils are derived from ancient basement rock and are mostly ultisols and alfisols. Outcrops of granitic basement rock form inselbergs or kopjes. The plant life is quite similar to that farther north in East Africa and is dominated by acacias and thickets of corkwood and other shrubs. In the south is a sharp boundary with the tropical dry forests that separate East African savannas from those of Southern Africa.

Southern African savannas. Separated from the East African savannas but with many of the same species or close relatives is another large area of savanna extending from east-central Namibia through Botswana and Zimbabwe into Mozambique and southward into South Africa, where it is called bushveld. These southern savannas stretch from 18° S to 34° S latitude across the southern end of the African Plateau. Elevations are 2,300–3,600 ft (700–1,100 m) above sea level. A single rainy season during the Southern Hemisphere summer is characteristic. Rains come with an easterly flow of moist air off the Indian Ocean that is drawn into the subcontinent when the ITCZ is displaced toward the Tropic of Capricorn. Rains usually take the form of afternoon and evening thunderstorms.

Most of the southern savanna region is associated with the vast area in Africa covered by Kalahari sands, the world's greatest deposit of sand, which stretches from the Orange River in South Africa to just north of the equator. The exact origin of these red sands is unknown, but they may have first accumulated nearly 3 million years ago. More recently in geologic time, during the alternating wet and dry periods of the Pleistocene Epoch, they were weathered and redistributed. Today, the dunes are stabilized by vegetation and are what some have called a fossil desert. In the north, the sands are covered by tropical forests, but in the Kalahari Depression, a dry savanna occurs that serves as a transition zone into the very dry Namib and Karoo deserts to the west.

A wetter type of savanna, much more similar to those of East Africa, lies north and east of the Kalahari across northern Namibia, Botswana, Zimbabwe, and Mozambique. Here in south-central Africa broad-leaved, thornless trees of *Brachystegia* and *Julbernardia* dominate. This vegetation is known as miombo. It grows on moist but infertile soils. Although most of the trees are legumes, few have root nodules housing bacteria that can fix nitrogen from the air. Higher plants must rely on the nitrogen-fixing capabilities of leguminous forbs, soil bacteria, and surface crusts of cyanobacteria. The mostly perennial grasses are not the upright bunchgrasses that are typical elsewhere, but they tend to lie prostrate along the ground and reproduce vegetatively by stolons.

Fire is a recurring event in these savannas, and trees and shrubs have the ability to resprout when damaged by fire or by elephants. Fire and large herbivores are important in keeping the vegetation open and preventing encroachment of closed woodlands. Grasses provide the fuel for savanna fires. When fires ignite at the end of the dry season, before the flush of new grass growth, tree seedlings are killed. Grazing animals and tree cover determine whether enough dry grass is available to support an effective burn. When average annual precipitation is greater than 25 in (650 mm), frequent fires are sufficient to keep the savanna vegetation open. With less than 25 in of rain a year, both fire and browsing animals are necessary to maintain an open canopy in the tree layer and thus an understory of grasses.

South of the Limpopo River, in the eastern part of the southern African subcontinent, miombo gives way to a mixed savanna containing numerous acacias. These fine-leaved trees and shrubs grow on fertile soils under a moist savanna climatic regime. Africa's first national park, Kruger National Park in northeastern South Africa, exemplifies this savanna type, actually a complex mosaic of plant communities determined largely by local geology—namely, the contrast between ancient granitic outcrops and younger volcanic bedrock.

Moist savannas are home to Africa's "big five": elephant, rhinoceros, buffalo, lion, and leopard. So-called megaherbivores—elephants, white or square-lipped rhinoceros, black or hook-lipped rhinoceros, hippopotamus, and buffalo, weighing more than a ton (1,000 kg) each—account for more than half of the biomass of large herbivores. Browsers—herbivores that consume primarily the leaves, twigs, shoots, fruits, and flowers of shrubs and trees—are major elements in the animal life in this type of savanna. Elephants, black rhino, eland, and kudu are among the larger browsers. Smaller browsers include hinged and leopard tortoises and abundant caterpillars and grasshoppers. Many large ground birds feed on the abundant insects. Most commonly seen are quail-like francolins.

Tall termitaria, the chimney-shaped mounds of termites, are conspicuous on the landscape. The watchful tourist may glimpse another insect important to nutrient cycling in the savanna, the dung beetle (see Figure 3.16). A brown golf-ball-size sphere may be spotted rolling across the road. Closer observation reveals it is being propelled by a large scarab beetle, the dung beetle or tumblebug. Many African dung beetles feed exclusively on the feces of the large herbivores that most tourists come to see. A pair will roll the dung into a ball and bury it in soft earth, whereupon they mate and deposit an egg in the center of the ball. As the larva develops, it feeds on the undigested plant material so abundant in the droppings of elephants, rhinoceros, buffalos, and other large mammals. Adults also eat dung, but they squeeze and suck from it the juices rich in microorganisms and other nutrients. Vast amounts of dung are removed from the surface and decomposed by the activities of these insects.

In the north, the savannas of southern Africa transition into dry tropical forests on the infertile soils developed on the ancient crystalline rocks of the Gondwanan Shield. The southern limits are reached at the higher elevations near the rim of the African Plateau where the Temperate Grassland Biome, South Africa's highveld (see Chapter 2), takes over.

Figure 3.16 A dung beetle rolling its ball of dung. *(Photo © Joy Stein/Shutterstock.)*

Dry savanna: The Kalahari. Often called the Kalahari Desert, the region is better designated as a dry savanna. It does share many species and genera with the Nama-Karoo shrublands, but it lacks most of the succulents so characteristic of those deserts because frosts are frequent and often severe. Indeed, grasses with an open upper story of shrubs and trees form the dominant vegetation (see Figure 3.17). The Kalahari is a simpler ecosystem than the true deserts to its west, with many fewer species overall. Its position as a transition between moist savanna and desert is reflected in its having few endemics. A sizable tract spanning the border region of Namibia and Botswana has been set aside as Kgalagadi Transfrontier Park, one of the world's largest national parks.

Low rolling sand hills (fossil dunes), nearly circular salt pans, and dry river beds make up the Kalahari's landscape. The rivers were once part of the Orange River drainage system, but they were blocked by dunes long ago. Today, they rarely have water flowing between their banks, but they remain valuable sources of salt and forage. The Kalahari experiences hot summers and cold winters. In the summer, air temperature can reach well above 100° F (40° C), but ground temperatures in exposed areas may rise to 160° F (70° C). Rainfall is low most years and unpredictable. Most rain falls during summer (November to April) thunderstorms. In Kgalagadi National Park, the annual average is about 10 in (250 mm), but actual total amounts vary from year to year between a recorded

Figure 3.17 Dry savanna of the Kalahari with springbok, an antelope well adapted to dry environments. *(Photo by author.)*

low of 2.2 in (56mm) and a high of 26 in (660 mm). Wet years happen every 10 to 20 years. Only then does water flow in the rivers and sometimes overflow their banks. In the Park, the Nossob River last flooded in 1963 and the Auob in 2000. In wet years, population explosions among annual grasses, forbs, and creepers and an extraordinary abundance of caterpillars and rodents provide a feast for herbivores and carnivores alike. The plant matter dries into a flammable mass that is ignited by lightning during the next spring's dry thunderstorms. Fire is thus a recurring factor in the park ecosystem. In the year or so after heavy rains, blue wildebeest, red hartebeest, and springbok may migrate southward more than 100 mi (200 km) from Botswana to South Africa. What looks like a mass migration is actually made up of numerous small herds averaging 80 animals. Today, the migration is halted by fences along the southern and western boundaries of the park.

Dry years with below-average precipitation are much more frequent than wet ones. During these years, animals such as gemsbok and red hartebeest depend on bulbs and rhizomes and compete more strongly on the grassy areas with other grazers such as springbok, blue wildebeest, and Ostrich. Many antelope migrate northward to wetter habitats; but gemsbok, well adapted to desert conditions, remain, as do Ostriches.

The grasses of this dry savanna are mostly annuals such as Kalahari sourgrass and nine-awned grass, but perennial tufted grasses are also important for grazing animals. White buffalograss and the tall, small, and silky bushman grasses are among the better forage plants. The more common Lehmann's lovegrass has limited grazing value. Certainly, the most important tree in the ecosystem is camel thorn, an acacia (see Figure 3.18). This is a widespread species found from Angola to Mozambique to Namibia, South Africa, Zambia, and Zimbabwe, and always growing on the Kalahari sands. It usually occurs as solitary or scattered individuals, rarely as dense woodland. Mature camel thorns may reach heights of 35 ft (10 m) and have taproots reaching depths of 130 ft (40 m). Their pods are rich in protein and important food for a variety of animals. The trees provide shade and shelter for lizards, birds, tree rats, and antelopes.

The untidy grass nests of the White-browed Sparrow-weaver and the dense thatched "apartment" dwellings of the Sociable Weaver may adorn the branches. Birds visiting the camel thorn rain down seeds in their droppings, so at the base of the trees, a community of raisin bush, buffalo thorn, annual bur bristle grass, and perennial guinea grasses develops. These become highly flammable during the dry season—as do the birds' nests—and many older trees succumb to fire.

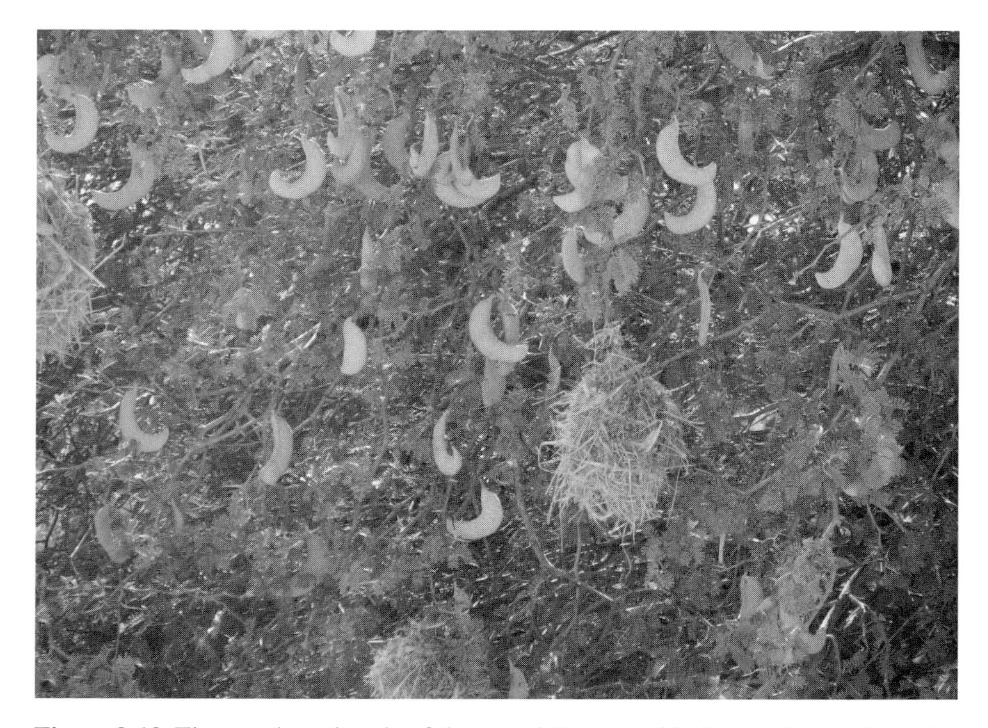

Figure 3.18 The ear-shaped pods of the camel thorn are highly nutritious. Note also the hanging nests of weaver birds. *(Photo by author.)*

Mammals attracted to the midday shade of camel thorn enrich the soil and create habitat for soft ticks. The pods that fall to the ground beneath the trees attract termites, which, in turn, are eaten by nesting birds. Beetles lay their eggs in the fallen pods. Interestingly, when the pods are consumed by antelope, the beetle larvae are killed, yet their presence actually helps in the germination process of the camel thorn's seeds because of the holes they have dug in the seed covering. Those seeds that have passed through the digestive tract of animals and are deposited in their dung germinate at much higher rates than those left uneaten. It may be that the holes allow more water to penetrate the seed.

Among other small trees of significance is the white-stemmed shepherd's tree. Its leathery, grey-green leaves are rich in vitamin A and browsed by springbok and kudu. Its dense shade draws mammals looking for relief from the summer heat, especially lions and leopards. Whereas in the sun the surface temperature may be above 160° F (71° C), under a shepherd's tree the ground may be only 70° F (21° C).

Numerous shrubs grow in the Kalahari. Blackthorn, a deciduous acacia, provides forage for smaller antelopes that consume its fallen leaves, pods, flowers, and new shoots. The gum that oozes from breaks in the stems is a favorite treat of the Kori Bustard, so much so that aboriginal San peoples learned to use the gum to trap the giant birds. In winter, rodents amass dead grasses for shelter at the base of the shrub. These nests become a fire hazard, but the blackthorn can resprout after a burn.

Another plant that deserves mention is the annual tsamma melon, ancestor of the watermelon (see Figure 3.19). Tsammas grow on wiry creepers and the melons are 90–95 percent water. They have little nutritional value, but are used by Ostrich, mice, ground squirrels, antelopes, and brown hyenas as sources of moisture. Unlike the domestic watermelon, the rind is hard and the pulp firm. The fruit may remain edible for two years. Yet another interesting plant, one with possible future significance, is one of the few stem succulents in the Kalahari, wild ghaap. This edible cactus-like plant is known to have appetite-suppressing properties and is of interest to pharmaceutical companies seeking antidotes to the world's increasing battle with obesity.

Animals of the Kalahari require adaptations to heat and drought, and many of the large mammals characteristic of East Africa or the moist savannas of southern Africa are absent. Most of the birds are migratory. So is the blue wildebeest, which must drink water every two or three days. Today, many of the wildebeests migration routes are cut off by fences, and they suffer more than the other antelopes in dry years. In contrast are the truly desert-adapted gemsbok and springbok, which are yearlong residents. Both are able to extract the water they require from their food and conserve body water by producing concentrated urine and dry feces. They feed mainly at night when the water content of plant material is highest.

Many animals are nocturnal. Springhares, porcupines, and the secretive aardvarks spend the daylight hours beneath ground in cool burrows. The lions of the

Figure 3.19 Tsamma melon, a source of moisture in the dry Kalahari. It or one of its close relatives is the probable ancestor of the watermelon. *(Photo by author.)*

Kalahari are also active mostly at night, sleeping under trees or cliff overhangs during the day. Lions get the water they need from the blood and moisture of their prey, but will drink when water is available. Other nocturnal predators include the bat-eared fox and the Cape fox, Africa's only true fox. The latter consumes beetles, termites, rodents, and lizards. Despite the number of animals moving around, the darkness of the Kalahari is eerily quiet. The only sound is the clicking of the common barking gecko and that is only heard for a short time after sundown.

Some small mammals are active during the day. Among them are the highly social meerkats and ground squirrels, the yellow mongoose, and whistling rats. The aardwolf, a hyena, is active when the termites it preys on are; so it is diurnal in winter and nocturnal in summer.

Mammalian predators are fairly diverse. Being highly opportunistic, black-backed jackals are active both day and night (see Figure 3.20). They hunt termites, mice, and springhares, consume tsamma melons and other fruit, and scavenge carrion at the kill sites of larger carnivores. Brown hyena are likewise generalists in their diet, although they are primarily scavengers and nocturnal. Their droppings are white from the calcium in the many bones they digest. The cats of the Kalahari include lions, leopards, cheetahs, caracals, and African wild cats, the last closely resembling a domestic cat.

Figure 3.20 The black-backed jackal is a widespread medium-size carnivore in the African savannas. *(Photo by author.)*

Moist Savannas of Southern Africa: Kruger National Park

As mentioned earlier, Kruger National Park contains a mosaic of plant communities determined by local geology, but rainfall amounts also play a role. Average annual precipitation varies by latitude. Northern Kruger receives less than 20 in (500 mm) a year, whereas the rest of the park receives more, between 20 and 28 in (500–720 mm). The dividing line occurs at the Oliphants River or roughly at the Tropic of Capricorn. The Park lies below 3,000 ft (1,000 m), but frost may be experienced several times a year.

North of the Oliphants River is a dense tree and shrub savanna composed almost exclusively of mopane. This is a broad-leaved plant with butterfly-shaped foliage that grows on dry but fertile soils. Periodic destruction by elephants results in different sizes and ages of mopane growing together in dense stands. A few other woody plants are found in the area; most noticeable and most spectacular are large, often-solitary specimens of baobab. Along permanent and seasonal rivers are fever-tree, Natal mahogany, and sycamore fig. Nyalas and waterbucks may forage beneath them.

South of the Oliphants River is a bush savanna that has a much richer tree and shrub flora than the mopane north of the river. Grasses are more abundant and support more herbivores and hence predators, so Africa's charismatic animals, large and small, are more frequently encountered by park visitors here than in the northern reaches of the Park.

Elephants are a keystone species, helping to reduce the cover of woody species as they tear branches off the trees, debark their trunks, and knock over or simply trample both trees and shrubs. While they are essential in keeping the grasslands open enough to make habitat for grazing animals, the elephant population easily becomes too large and their activities can become too destructive. Park managers must struggle with ways to keep their numbers in check.

The diverse plant life allows for a diverse fauna. The animals of these southern savannas are much the same as those on the East African savannas. The most common browser is the impala, although kudu and giraffes are also plentiful. Plains or Burchell's zebra and impalas are abundant in the grassy areas (see Figure 3.21). Guineafowl and francolins consume young shoots of plants, seeds, and insects. Go-away birds (lories) are fruit eaters, but they also like flowers and flower buds. Weaver-finches eat seeds and insects and dangle their nests from the tips of tree branches. In Kruger, unlike the dry Kalahari, there are deep permanent pools in rivers and ponds, which are habitats for crocodile and hippopotamus, animals not found in the drier parts of southern Africa.

Figure 3.21 Zebra and giraffe are just two of many large mammals that feed at different heights and on different plants in the African savannas. *(Photo by author.)*

Ecotourism is important to the economy of South Africa, and Kruger National Park is one of Africa's oldest protected areas. Nonetheless, poaching continues to be a threat as does the collection of firewood by nearby villagers. Controlling the illegal killing of elephants has been successful, but it has created new problems as confined elephant populations have grown beyond the Park's carrying capacity. Much of the land surrounding the Park is now farmland, grazing land, or land converted for urban uses, including those directed at the tourist, with very little, if any, buffer zone. The loss of natural habitat is not the only problem. Development comes with its own set of negative impacts, such as the killing of predators and scavengers, use of pesticides, and introduction of nonnative species. The management of large tracts of savanna as natural areas thus includes balancing the needs of the wildlife in a diminished habitat with those of the visitor eager to see Africa's big animals.

Australian Savannas

Savannas occur across northern Australia, curving southward along the east coast of Queensland into the subtropics (see Figure 3.22). They are located between 17° S and 29° S, in a region influenced by monsoonal climate patterns. Rainfall ranges from 60 in (1,500 mm) a year in the north to about 20 in (500 mm) in the southern part of the savanna region. Most rainfall occurs during the Southern Hemisphere's summer months of December through March.

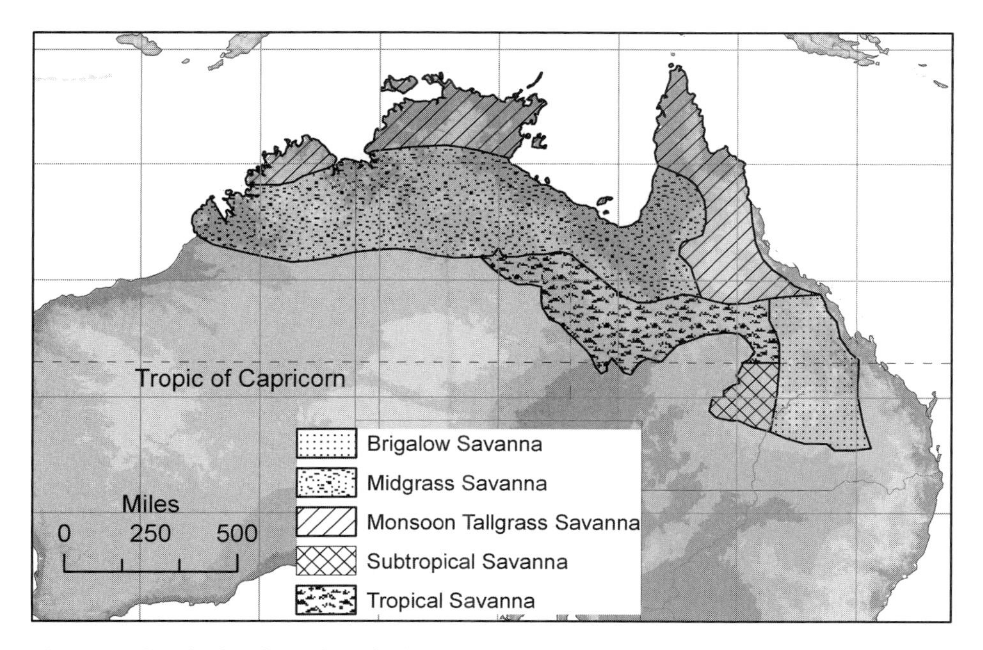

Figure 3.22 Distribution of tropical savanna in Australia. *(Map by Bernd Kuennecke.)*

Australians recognize at least six major types of savanna that differ from each other because of rainfall amounts and soil conditions. Most have tree and shrub layers dominated by gums (eucalypts) and acacias. Other woody plants include paperbarks, banskias, terminalias, and bean trees. Grasses are C_4 bunchgrasses.

The Monsoon Tallgrass Savanna forms the northernmost zone, where annual rainfall amounts to 30–55 in (750–1,400 mm). It has developed on low-fertility soils. The major trees are gum trees. Common grasses are red oat grass and annual sorghums. Tropical Tallgrass Savanna (also known as black spear grass savannas) occurs south of 21° S along the east coast and grades into a Subtropical Tallgrass Savanna south of the Tropic of Capricorn. The black spear grasses and kangaroo grasses dominate both aspects.

Midgrass Savannas with three-awns dominating are found on poor soils in the Northern Territory and Western Australia. Other midgrass communities exist in Western Australia on black cracking soils. Finally, Mulga Pastures are found throughout northern Australia on poor, dry, stony soils. Mulga is a type of acacia.

In addition to these six types of savannas—influenced by fire, but considered natural, are derived savannas in southern Queensland, created by the burning of Brigalow (*Acacia harpophylla*) woodlands. Human-set fires have eliminated the undergrowth, leaving scattered brigalow and bottle trees (*Brachychiton rupestris*).

Australia's animal life is like no other continent's. Diversity is not great, but most species are unique. Among mammals, marsupials stand out and this is certainly the case in the savannas. The largest marsupial in the biome is the grey kangaroo. Others include wallabies, walleroos, bandicoots, and quolls. Brushtail possums and sugar gliders are the counterparts of tree-living rodents in other parts of the world. Several rats are some of the few placental mammals in Australian savannas.

Many kinds of birds inhabit the Australian savannas. Members of the parrot family such as lorikeets, cockatoos, and parrots themselves are abundant. Other bird families represented include the honeyeaters, butcherbirds, and finches. The Emu is the signature flightless bird of Australian grasslands. Reptiles include frill-necked lizards, goannas, dragons, skinks, and snakes.

South American Savannas

Two major savanna regions occur in South America (see Figure 3.23). In the Northern Hemisphere, north of the Amazon rainforest in the Orinoco River basin, are the grass savannas of the Llanos; and south of that vast forest, in the Southern Hemisphere, are the savanna woodlands of the Brazilian Highlands known as cerrados. Although both of these occur in the tropical wet and dry climate typical of savannas around the world, their character is ultimately controlled by edaphic conditions—that is, seasonal flooding in the case of the Llanos and heavily leached

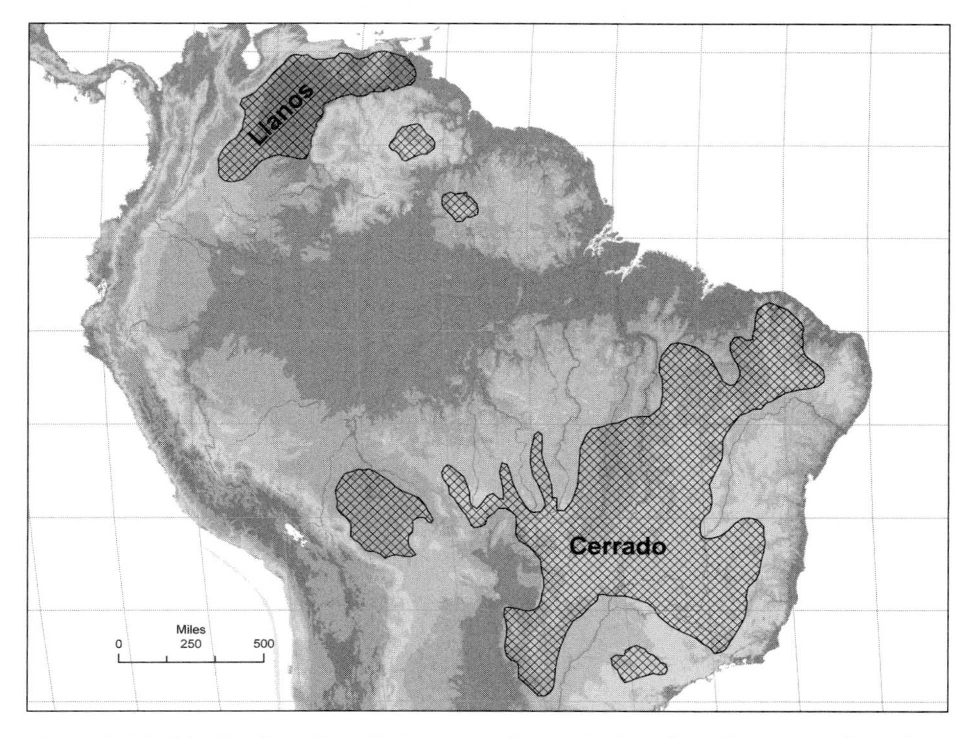

Figure 3.23 Distribution of tropical savanna in South America. The two main regions are the llanos north of the Amazon River and the cerrado on the Brazilian Highlands south of the Amazon. *(Map by Bernd Kuennecke.)*

soils in the case of the cerrados. They each present a different set of habitat conditions for both plants and animals.

Other places in tropical South America where savanna can be found are relatively minor on a global scale and will not be described here. They include the Rupununi-Roraima Savannas on the ancient surfaces of the Guiana Highlands in southern Venezuela and Guyana and in northern Brazil, and the seasonally inundated Llanos de Mojos in the Bolivian Amazon. The Pantanal, a vast area of wetlands and periodically flooded savanna woodland in the headwaters of the Paraguay River in southern Brazil, is considered here to be part of the Freshwater Biome.

The Llanos

The term *llanos* is Spanish for plains, but as a proper name, it refers to the expanse of savanna on the left (north and west) side of the Orinoco River in Colombia and Venezuela. The land tilts generally toward the east and lies at elevations ranging from 1,600 ft (500 m) in the interior to 300 ft (100 m) near the river's delta. Being between the latitudes of roughly 5° N and 8° N, these lowlands experience the

The Vanished Beasts of South American Grasslands

Few large mammals roam the temperate or tropical grasslands of South America today, but this was not always the case. During the Pleistocene and up to about 10,000 years ago, a fauna of giants was present. The variety of megaherbivores now known only from fossilized remains gives credence to theories that Neotropical grasslands are indeed ancient and were once even more widespread than now.

Glyptodonts, creatures the size of a small economy car, were covered in a bony shell like their extant relatives, armadillos, and had hoof-like claws. They probably consumed pampas grasses. Toxodonts also roamed Argentina's ice-age pampas. Members of an endemic South American order of hoofed mammals, the Notogulata, the toxodonts stood 5 ft (1.5 m) tall at the shoulders and resembled rhinoceroses. Their curved upper-cheek teeth keep growing throughout their lives, suggesting an adaptation to a diet of abrasive grass.

Gomphotheres were elephants with four tusks. They arrived from North America after the closing of the Isthmus of Panama and became abundant during the Pleistocene. The most common, *Stegomastodon*, had teeth adapted to grinding fine grasses.

Other large plant eaters included ancient camels, some giant ground sloths, giant capybaras, and the last of the endemic order Litopterna, the three-toed *Macrauchenia patachonia*. The litoptern looked like a llama with a short trunk. (The first fossil specimen was found by Charles Darwin during his investigation of Patagonia in 1834.) These animals and many others were grazers, browsers, or both and would have thrived in savanna-like vegetation. Almost all had disappeared by 10,000 years ago. Among the survivors are smaller animals that range through both humid and arid biomes in South America, mammals such as tapirs, peccaries, and armadillos. Grassland specialists are no more.

typical tropical temperature pattern of year-round warmth and little variation from month to month. Closeness to the equator provides for a lengthy period of influence from the ITCZ and a single rainy season lasting some seven months, between April and November. Total precipitation can be close to 100 in (2,500 mm) a year in the southwestern parts of the region, much more than usually expected in the Tropical Savanna Biome. Toward the eastern edges of the Llanos, precipitation drops to 30 in (800 mm) a year.

Much of the surface on which the Llanos occurs is underlain by lateritic hardpan that impedes drainage and results in standing water for months at a time. During the dry season, the water evaporates, leaving a hard surface crust. Few plants can tolerate the alternating conditions of flood and drought. Among those that can are certain grasses and rushes and some kinds of palm. Poor drainage is a key cause of savanna in this part of South America. Four distinct vegetation subregions exist in the Llanos, reflecting drainage differences (see Figure 3.24).

The Overflow Plains cover 20 percent of the area. Several habitat types are included in this designation. All are submerged beneath flood waters for part of the year. The riverbanks themselves are well drained and support gallery forests. The

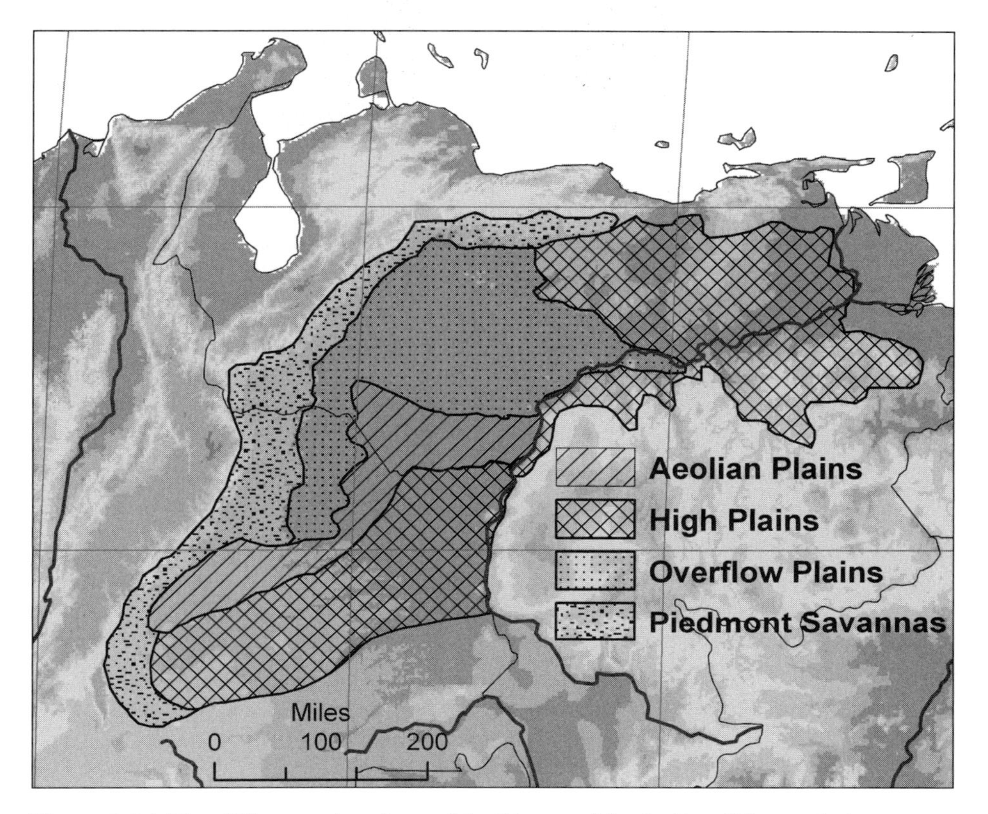

Figure 3.24 The different subregions of the Llanos. *(Map by Bernd Kuennecke.)*

banks may be 3–6 ft (1–2 m) above the streams at low water, but they will be underwater at the wettest times of year. About half of the Overflow Plains are composed of low basins (*bajiós*) covered with water at times, but dried out completely in November or December. The soils in these basins have a high clay content, which contributes to poor drainage and standing water during the rainy season. Moriche palms tolerate these conditions and grow in groves (*morichales*). Their fruits are food for wildlife and people, and their fronds are used for thatched roofing material and fiber for hammocks and cloth. Thousands of low mounds are covered with *Trachypogon* grasses. Spiny shrubs and a number of other grasses also grow in the bajiós. A third habitat type in the Overflow Plains are *esteros.* These develop in the lowest areas where fine sediments have collected. Drainage is so poor that most water is lost through evaporation rather than by percolation into the soil. As a consequence, esteros dry up only at the very end of the dry season (March or early April). Floating vegetation such as water hyacinth dominates, but plants rooted into the ground such as sedges and tropical morning glory vines are common.

The Aeolian Plains (aeolian refers to deposits of wind-blown sands or silts) formed on old sand dunes that originated during dry periods in the Pleistocene.

Sandy soils are both low in nutrients and excessively drained, so that even in humid climates they can create dry environments. Such conditions are rarely tolerated by trees. The Aeolian Plains are covered with poor-quality grasses. Moriche palms may form gallery forests along waterways and provide much of the food available for animals in this subregion.

High Plains occur in the extreme south of the Llanos and again east of the Overflow Plains. The topsoils have eroded away, exposing laterite hardpan that forms tablelands or mesas. Tree roots cannot penetrate the laterite and so trees are largely absent. Soils are typical oxisols: acidic and rich in iron and aluminum compounds. The trachypogon grasses that grow in these conditions are low in nutritional value. Floodplains that are close to streams and seasonally are under water support woodlands of either saladillo or chaparro manteco. The latter produces a high-fat fruit (*manteco* means lard) important to white-tailed deer. Its thick bark is high in the tannic acids used locally for tanning leather. Another small tree of the floodplain is the sandpaper tree, which concentrates silica in its leaves, making them coarse and rough. People actually use the leaves as sandpaper.

The fourth subregion is that of the Piedmont Savannas. These lie near the Andes Mountains along the northern and western edges of the Llanos. Soils are derived from sediments recently brought from the mountains and tend to be deeper and richer than elsewhere in the Llanos. Precipitation is high, generally 40–80 in (1,000–2,000 mm) a year. Almost all of it falls between April and November, and the savanna is flooded from July into October. January to April is the dry season, when most bodies of water dry up.

Large mammals are relatively scarce in the Llanos, especially in comparison to the East African savannas. White-tailed deer, the same species familiar to many residents of the United States and Canada, live on these tropical grasslands. A strange and unfamiliar mammal living there with deer is the giant anteater, the New World's counterpart to Africa's termite- and ant-eating aardvarks and pangolins. Anteaters are members of the order Xenarthra, which is nearly endemic to South and Central America and which also includes the sloths and armadillos. Carnivores venturing onto the Llanos include puma, jaguar, and ocelot.

Rivers, gallery forests, and morichales are important to the survival of a wide range of species that would not occur in their absence. The rivers support not only side-necked turtles and caiman and the largest snake in the world, the anaconda, but giant otters and crab-eating foxes. The Llanos are home to the world's largest rodent, the semiaquatic capybara. Mature specimens weigh in at around 175 lbs (79 kg). They are a bit like miniature hippopotamuses, living in family groups and spending much of their time in the water. The gallery forests are home to a number of New World monkeys that perhaps are more readily associated with tropical rainforests, such as howler monkeys, squirrel monkeys, and capuchins.

In terms of animal life, South America is really known for its great diversity of birds; and the Llanos is no exception. Some 475 resident and migratory species have been reported. The importance of water is shown by the fact that many are

Capybara: A Better Cow?

The name of the world's largest living rodent, capybara, comes from the Tupi-Guaraní language of South America and translates as "lord of the grasses," or, perhaps, merely "eater of grasses." Capybaras are stocky animals about 20 in (50 cm) tall, weighing as much as 120 lbs (50 kg). They graze grasses, but like Africa's hippopotamuses, spend much of their time in water. Their feet are partially webbed, and they swim with ease. When frightened, capybaras race into the water and hide submerged among aquatic plants. Only their nostrils, placed near the top of the head, may break the water's surface.

Capybaras congregate in small herds. Most of the land they occupy on the Llanos is cattle range. Ranchers view capybaras as competitors for the grasses their poorly suited livestock need. Illegally harvested for their meat and leather by subsistence hunters and small-scale commercial operations for many years and exterminated by ranchers as pests, capybara numbers declined in some areas, although they were never threatened with extinction. Nonetheless, conservation of the capybara is seen as a way to conserve wetlands. Particularly in Venezuela, capybara ranching could be a viable, legal, and sustainable way to both manage the rodent herds and protect their habitat.

Capybara convert grass to meat nearly three times faster than do cows, grow faster, and have a reproductive rate six times higher than cattle. The meat is tasty. However, the main product is high-quality leather made from their hides. Exported under the name "carpincho," the leather is unique in that it stretches in one direction only and makes excellent gloves as well as belts, handbags, and jackets. This is a mammal that could be conserved along with its native habitat by exploiting it. Raising capybara may make more sense than raising cows. Some ranchers are convinced of this fact, but markets for products must be developed if they are to be successful.

shorebirds such as ibises, storks, herons, and egrets. Frogs and toads also are abundant because of the abundance of seasonal and year-round water bodies.

Much of the Llanos remains relatively undisturbed at present, but both Colombia and Venezuela have plans to develop the region. Cattle raising is the primary economic activity today and in several ways threatens the long-term survival of the natural environment. An increase in the number of fires—used to stimulate the growth of new shoots for the cattle to eat—can change the composition of the vegetation. Exotic grasses, better adapted to grazing than native South American ones, have been introduced. These include Bermuda grass, fingergrass, and elephant grass from the African savannas.

Ranchers usually try to eliminate wild species that compete with cattle for grass or that may be pests or predators of livestock. In the Llanos, however, progress has been made in conserving two such species, the capybara and the spectacled caiman, by making these natives commercially important. Since the capybara is an aquatic animal, the Catholic Church approved it as a substitute for fish on Fridays, when followers of the faith should avoid consumption of poultry and meat. This has created a demand for capybara flesh and encouraged the management of their populations for sustained, commercial production. Caiman are farmed because their hides, made into leather, are valuable.

Transportation systems are vital to the health of a commercial economy, but they can spell problems for natural habitats. New highways across the Llanos run in a north-south direction toward major ports. This is perpendicular to the flow of water over the surface and could block or divert natural waterways. Similarly, proposed improvement of the Apure River for navigation in the western Llanos of Venezuela could alter the all-important water-based ecology of the region.

Cerrado

The largest area of tropical savanna in South America occurs on the oldest surfaces of the Brazilian Shield. Since almost all of this area lies in the Portuguese-speaking country of Brazil, the vegetation is universally referred to as *cerrado*, a Portuguese word that means "closed" and may have originally referred to the difficulty of riding a horse across the shrubby landscape. All of the general types of savanna (see Table 3.1) are found, but in Brazil, they go by a different set of names (see Table 3.3 and Figure 3.25).

Together these tropical grasslands and open woodlands grow as a complex patchwork of plant communities that originally covered more than 750,000 mi^2 (2,000,000 km^2) or about 22 percent of Brazil's land area. They stretch from the equator south to the Tropic of Capricorn at elevations that range from 1,000 ft (300 m) to 6,000 ft (1,800 m) above sea level. In the west and northwest, the cerrados border the Amazon rainforest; in the northeast, they abut the dry shrublands known as *caatinga*; and in the south and southeast, they give way to the tropical and subtropical evergreen forests of the Atlantic Forest. In the southwest, the world's largest tropical freshwater wetland, the Pantanal, is located.

The pattern of the patchwork landscape seems largely related to differences in soil and water availability; yet, fire frequency or geomorphic cycles of landscape

Table 3.3 Types of Cerrado, Brazil

Vegetation Type	Description
Campo limpo ("clean field")	Grass savanna: grasses 12–20 in (40–50 cm) tall; no trees or shrubs
Campo sujo ("dirty field")	Grassland savanna with some widely scattered shrubs and small trees
Campo cerrado ("closed field")	Tree savanna: primarily a grassland, but with scattered trees and shrubs
Cerrado *sensu stricto*	Savanna woodland: trees 10–25 ft (3–8 m) tall are the dominant element of landscape, but a continuous grass cover is present
Cerradão	Woodland savanna: almost closed woodland with trees 25–40 ft (8–12 m) tall

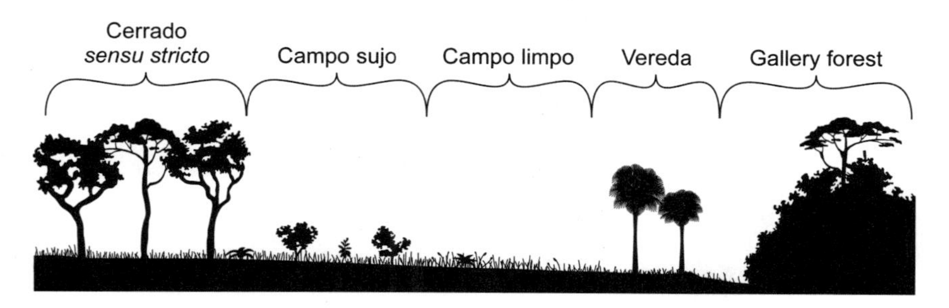

Figure 3.25 Vegetation profile showing the several types of cerrado as well as the location of the palm-dominated *veredas* and the dense gallery forests edging streams. *(Illustration by Jeff Dixon.)*

development also may play a role. For tree seedlings to survive the dry season, they must be able to extend their roots to depths below the grass root zone where they will not have to complete for soil moisture. They also must build up underground reserves of energy to resprout after a disturbance such as fire. Frequent fires will favor grasses over trees lacking the capability to resprout. When fires are prevented, woody plants of the cerradão invade and convert the savanna into a closed or nearly closed woodland.

The aerial parts of trees grow slowly, because energy is first distributed to the root systems. Plant growth is slowed by the low nutrient content of highly leached soils derived from ancient bedrock. Plants associated with tropical dry and humid forests do not flourish in such conditions. Erosion of the upland surfaces exposes less-weathered, less-leached rock residues, however, and in the southern parts of the biome, the Atlantic Forest encroaches upon the younger soils of valley bottoms. In the drier northeast, *caatinga* (a shrubland vegetation) is expanding at the expense of cerrado on younger surfaces (see Plate XV). In a geologic time frame, it seems that cerrado is an old assemblage of plants that is fated to shrink in area coverage as the ancient surfaces on which it occurs are worn away.

As an old group of plants, cerrado should be expected to harbor a large number of different kinds of woody plants and herbs, and this is indeed the case. The variety of higher plants found in the cerrados is second only to the diversity of plants in the tropical forests in South America. An estimated 10,000 species occur in the cerrados, the highest plant diversity of all the world's savannas. Scientists have identified at least 800 different kinds of trees and shrubs and many more grasses, forbs, and subshrubs. A given site may not contain a high number of species, but the species composition changes over rather short distances so the overall total (or what ecologists call beta diversity) is great. Diversity is highest in the center of the region. Since a large proportion of the plants—some 44 percent—are endemic to the region, the cerrados are one of South America's most species-rich biomes (see Table 3.4). The South American cerrados are considered to be one of the world's 25 critical hotspots of biodiversity.

Table 3.4 South America's Five Most Species-Rich Biomes

	Location	Biome
1	Amazon Forest	Tropical Rainforest
2	Atlantic Forest	Tropical Rainforest
3	Cerrado	Tropical Savanna
4	Caatinga	Tropical Dry Forest
5	Pantanal	Freshwater Wetland

The Brazilian Highlands are a vast area of exposed continental shield. At least three distinct surfaces appear as step-like plateaus separated from each other by long, steep escarpments (see Plate XV). The bedrock consists of acidic shales and crystalline rocks that have been deeply weathered over millions of years and leached of most soluble compounds. Infertile red and yellow oxisols (or latosols) are the main types of soil present. Most of these soils are deep and well drained. In some locations, aluminum has concentrated to levels that are lethal to many crop plants, but native cerrado species are able to tolerate this toxic soil chemistry.

On the gently undulating plateau surfaces surrounding the headwaters of streams, the water table rises to the surface during the rainy season and saturates the soils. Since most cerrado plants do not withstand waterlogging in the root zone, a distinct swamp community of grasses and palms mark these valley-side wetlands, which are called *veredas* (see Figure 3.25). Other seasonally saturated soils occur in low-lying areas flooded each year by the many rivers that cross the region. Riparian or gallery forests trace the courses of these streams, headwaters, and links to major rivers such the Amazon, Paraguay, São Francisco, Araguaia, and Tocantins.

In the tropical wet and dry climate region of this part of South America, 30–80 in (800–2,000 mm) of rainfall are received each year during the six- to seven-month-long rainy season. It is usually very dry from April through September. Temperatures are normally warm throughout the year, though seasonal differences become more pronounced close to the Tropic. Occasional outbreaks of cold air from the south can bring frost to much of the region in some years.

The trees of the cerrado are quite distinctive in appearance and for the most part are unlike those of African and Australian savannas. Trees tend to have either thick, tough, and large leaves or finely divided compound leaves. Examples of the former are pequi and murici; examples of the latter include legumes. On many large-leaved trees, the veins on the undersurfaces of the leaves are conspicuously raised. In lixeiro, silica is deposited in tissue on the undersides creating a sandpaper-like surface. The leaves of other species, such as *pau santo* and *pau de arara*, are arranged as a rosettes at the tips of branches.

Although a few woody plants are truly evergreen, most are deciduous. The deciduous trees, however, keep their leaves well into the dry season and therefore must be adapted to drought. Some may be completely bare for only a few days (brevideciduous). Others replace only some of their leaves each year (semideciduous). Either way, new leaves grow before the summer rains begin. Large leaves are waterproofed by a thick cuticle. As a result, many broad-leaved trees have a similar gray-green color. Sunken stomata may further limit water loss.

The subshrub is a distinctive and common growthform in cerrados. They typically have swollen organs below ground (rhizomes, corms, or xylopodia) that store energy and nutrients. Their aerial stems die during the dry season, but several have curious underground systems of woody structures that make them essentially underground trees. Cabbage bark, catuaba, and monkey-nut are examples of this growthform. The new shoots that sprout from these subterranean "branches" may cover an area 35 or more feet (10–12 m) in diameter. What looks like a cluster of individual plants on the surface in the wet season is actually the many "twigs" of a single plant.

The most commonly occurring trees have thin, spindly trunks and stand only 10–25 ft (3–8 m) tall (see Figure 3.26). The trunks and branches are twisted and bent and have thick, corky bark, often deeply grooved. Such a growth pattern may be the result of the low fertility and high aluminum content of soils.

The leaf structure in the broad-leaved trees may be a response to soil conditions rather than drought. High amounts of aluminum reduce the solubility of two elements important to normal plant growth, calcium and phosphorus, and thereby deprive plants of adequate amounts. Without key nutrients, the plants tend to produce more cellulose, more of the structural cells of leaves (sclerenchyma), and

Figure 3.26 Cerrado *sensu stricto* is an open stand of small twisted trees with the complete cover of grasses that makes it a savanna. *(Photo by author.)*

more cuticle. Many woody plants of the cerrado have special ways to keep aluminum from being absorbed by the roots or—when aluminum compounds are drawn up into the plant—ways to concentrate it and store it where it cannot damage the plant. It is often stored in the midrib and other tissues of the leaves. When the leaves are shed, the plants get rid of the excess aluminum. The *pau terra* trees, endemic to the cerrado, are examples of plants that exclude aluminum. Indeed, some can grow only where there is a high concentration of aluminum in the soils.

Drought apparently frequently kills or damages the buds at the tips of branches at the end of the growing season, effectively pruning the trees. Much like a tree damaged in an ice storm in northern forests, side buds take over and form a new main stems for the plant, so the tree grows crookedly. Tall, straight trunks are rare.

While contorted trees and strange leaf shapes give a distinctive look to the cerrados, it is the continuous cover of grasses that makes them savannas. The grasses are mostly perennial C_4 bunchgrasses. Their growth begins only after the first rains. Perennial forbs include members of the legume, sunflower, and orchid families. Annuals are notably rare. The legumes, of which more than 500 species have been reported, have symbiotic relationships with *Rhizobium* bacteria that fix nitrogen from the air and release plant-usable forms to the host and into the soil. Other plants, including terrestrial orchids, are associated with mycorrhizal fungi that enable them to obtain needed phosphorus. Grasses and forbs contribute much to the high plant diversity of the cerrado: some 4,700 species have been recorded. Most species of the herb layer are able to withstand ground fires. Their seeds, rhizomes, and xylopodia are protectively covered by soil. Fire serves to release nutrients that are otherwise bound into living and dead plant matter and apply them to the soil as a coating of ash. Herbaceous plants sprout or resprout within a few days of a burn; and grasses, terrestrial orchids, and others will flower profusely in a matter of weeks.

Epiphytes or air plants are a significant component of the vegetation of cerrado *sensu stricto* and cerradão, but they are nowhere near as important as in tropical dry and rain forests. Among the plant groups that grow with their roots dangling in the air are orchids, bromeliads, and *Rhipsalis* cacti. Vines are also rather diverse and include members of the morning glory, yam, and passionfruit families.

Those areas where the soils are waterlogged during the wet season do not have typical cerrado plants growing on them. Veredas have palm savannas dominated by buriti palm trees. Other types of palm grow in seasonally flooded areas on the floodplains of rivers. In some of these regularly inundated areas are a large number of mounds of uncertain origins. Many seem to be abandoned termite mounds. They form islands of better drained and more fertile soil during the rains and host woody plants that are unable to grow where there is standing water. These distinctive habitats are called *campos murundus* (earthmound fields).

Gallery forests are another important habitat. These dense thickets of trees and shrubs line the banks of streams (see Figure 3.25). Much of the rich animal life of the cerrado depends on these ribbons of forest for shelter, nesting sites, and food.

Plants of the cerrado have importance for local people and potential usefulness for the world's populations. More than 100 shrubs, subshrubs, and forbs have known medicinal value. Local residents use roots, bark, eaves, fruits, and seeds, which often are prepared as teas or bottled in alcohol. The stems and leaves of two plants, both called *arnica*, are sold throughout central Brazil for use as antibiotics. The roots and xylopodia of catuaba are used as aphrodisiacs, while those of *rabo-de-tatu* are remedies for stomach ailments.

A number of cerrado plants are related to important crop species and as such represent genetic resources for future crop improvement. For instance, in the cerrado, there are 42 wild species of *Manihot*, the genus containing manioc (or cassava), long a mainstay of the diet of inhabitants of the humid tropics around the world. Yam, sweet potato, cashew, vanilla, and pineapple are among other crops with wild relatives in this biome.

Animal life in the cerrado is quite different from that of the African savanna. Large mammals—those weighing more than 110 lbs (50 kg), so spectacularly abundant and diverse in eastern and southern Africa, are rare in the cerrados of Brazil. Among herbivores, tapir, marsh deer, and the largest specimens of giant armadillo qualify; among carnivores, there are only jaguars and pumas. In the geologic past, many others may have evolved as cerrado plants came to prominence, but they were lost when the isthmus of Central America formed at the end of the Pliocene and allowed the exchange of animals between North and South America or during the wave of extinctions that marked the late Pleistocene. The few large mammals that currently inhabit the region are found in many other South American biomes as well and are by no means restricted to the cerrado, although the maned wolf (see Figure 3.27) is primarily associated with the cerrado. Small mammals weighing less than 11 lbs (5 kg) account for 88 percent of the mammals, and most species are widespread rather than restricted to this one biome. Only a few mice are endemic. Rodents are the chief vertebrate grazers of grasses, a role performed by ungulates in Africa. Among the small mammals are peculiarly neotropical forms such as opossums, tamanduas, and armadillos, and large rodents such as agouti and paca.

Birds are especially diverse, with more than 800 species reported. Thirty percent are endemic. The most species-rich family by far is that of the New World or tyrannid flycatchers (Tyrannidae), with 111 species. Most conspicuous are ground-dwelling birds such as the rheas, the Red-legged Seriema, the Campo Flicker, and a number of partridge-like tinamous, which may share an early ancestor with the great ratites of the world.

The lowest diversity of birds occurs on the campo limpo or grass savannas, because many depend on gallery forests, patches of dry forest, and palm groves for food, nesting sites, and shelter. The world's largest parrot, the rare Hyacinth Macaw, is one such bird. It and other parrots, such as the Blue-fronted Amazon, consume 90 percent of the fruits of palms. They are sloppy eaters, however, and let leftover pulp and seeds drop to the ground, where it is devoured by opossums, rodents, tapirs, and maned wolves.

Figure 3.27 Maned wolf, the flagship species of the cerrado. *(Photo © Tim Grootkerk/ Shutterstock.)*

The Maned Wolf: A Flagship Species

Some animals capture the imagination of people more than others, and so people are moved to preserve them in the wild. Conservation groups single out such species to serve as symbols or ambassadors of a vulnerable habitat in the hope that they will draw attention to an environmental problem and help raise funds to overcome the loss or significant alteration of threatened ecosystems. Such species are called flagship species. Often they are large mammals with fairly large territories or home ranges, whose preservation requires the conservation of large areas of favorable habitat. Saving the flagship species means many less charismatic species will be protected as well. The maned wolf, the largest member of the dog family in South America, is a flagship species for the cerrado.

At first glance, the maned wolf looks like a large-eared red fox on long legs. It stands almost 4 ft (110 cm) high at the shoulders and weighs about 55 lbs (25 kg). Its orange-brown coat is set off by black legs and a black muzzle. When it senses danger, it raises black hairs on the back of its neck—its mane. It is fox-like in behavior, but it is neither a fox nor a wolf, but rather a unique South American canid.

Solitary, nocturnal hunters, maned wolves eat pacas, agoutis, and smaller rodents, as well as insects, reptiles, and birds; the mainstay of their diet, however, is fruit.

Maned wolves and the rest of cerrado wildlife are threatened by habitat loss as agriculture expands on the Brazilian Highlands. Some local people shoot maned wolves because they prey on domestic poultry; others do so because they believe certain wolf body parts have magical or medicinal properties.

The cerrados have many more tree-dwelling birds and many fewer ground dwellers than African or Australian savannas. Fruit-eating parrots (33 species) and the numerous nectar-eating hummingbirds (36 species) are just some of the birds. Seventy percent of breeding birds nest in gallery forests or remnant patches of dry forest, mostly during the rainy season. Others seek refuge in the forests during the dry season or grass fires and flee to them to escape predators. Both open country and arboreal species are important pollinators and seed dispersers for cerrado plants. A few are attracted to the grass fires that regularly sweep across the landscape. Catching insects and other small animals fleeing the flames are caracaras, seriemas, rheas, and the very abundant White-collared Swifts.

Reptiles include a greater proportion of endemics than is true for birds and mammals, and a number of endangered species. There are 10 tortoises and turtles, 5 crocodilians, 15 worm lizards, 47 lizards, and 10 snakes. The array of habitats in the cerrado landscape accounts, in part, for this high diversity. Several reptiles use underground dens. Others occupy termite and ant nests or armadillo burrows, while the giant worm lizard lives beneath the nests of leaf-cutter ants. Access to underground refuges can be a matter of life or death during grass fires.

The abundance of trees and shrubs has allowed some lizards and snakes to become arboreal and find shelter in tree cavities. The false chameleon is so well adapted to life in the trees that it has a prehensile tail.

The millions and millions of termites and ants are the main processors of plant matter in the cerrado. While huge towering mounds are not a feature of New World savannas, small mounds are abundant and common features of the landscape. Aboveground nests may be placed in tree branches or near the bases of trunks or on fence posts—seemingly wherever the mass of carton can be attached. Both types of insect live in highly organized colonies with populations that may number in the millions. It is said that a single colony is the equivalent of one cow in terms of the vegetable matter it consumes each day.

The cerrados are under extreme threat today, and many of the problems stem from agriculture and related developments in transportation and irrigation. The deep soils and level surfaces of the Brazilian Highlands encouraged modern, mechanized forms of commercial agriculture once techniques were developed to overcome the low-fertility and high-aluminum content of the soils and the prolonged dry season. The use of lime and chemical fertilizers helped improve the soils, and agricultural research stations developed high-yielding, drought-resistant varieties of soybean, rice, and corn. Cattle raising prospered when African grasses were planted in improved, artificial pastures. In addition, plantations of Australian gum trees (eucalypts) and North American pines sprung up to produce pulpwood. By the 1990s, two-thirds of the cerrado had been converted to cropland, tree plantation, or grazing land; by 2000, the proportion had risen to 80 percent. The last, most remote areas in the Tocantins watershed have recently been opened to settlement and exploitation with the

Ants Everywhere

Ants are crawling all over the place in the cerrado. They run up and down the trunks of trees and shrubs and busily inspect the leaves. Close relationships have evolved not only between ants and plants, but between ants and other insects. Pequi is one of the many small trees that has special glands on sepals or at leaf axils—extrafloral nectaries—that secrete a sweet substance and attract at least 34 different kinds of nectar-gathering ants. The ants defend these nectaries against browsing animals and thereby protect a plant's foliage from damage and removal. Plants with ants grow better than plants without them.

Other insects play the same game. Both the adults and young of some treehoppers, aphids, and scales (Homoptera) and the caterpillars of gossamer-winged (lycaenid) butterflies produce sweet droplets of a substance called honeydew that is rich in sugars and proteins and used as food by 21 different ant species. Honeydew gathering ants guard the insects as though they were their sheep or cattle. They kill or remove from the leaf any potential predators or competitors of the larvae and thus increase their chances of survival. The honeydew producers may actually let ants know where they are by flicking drops of honeydew onto the ground.

Some ants consume foliage. Leaf-cutter ants are a neotropical phenomenon and one of the major consumers of plant matter in the New World rainforests and savannas. The ants themselves do not eat the leaves but cut off pieces and carry them to their underground nests. A parade of leaf bits held aloft by small ants marching in single file is a common sight throughout the humid tropics of the Americas. In the nest, a series of ever smaller ants cut the pieces into tinier and tinier fragments until they make a mushy medium on which fungi grow. The ants farm and then consume the fungi, in ways similar to termites in the Old World savannas.

While leaf-cutter ants have their nests below ground, other ants colonize trees and shrubs. Different species build their nests of carton, chewed wood mixed with saliva, at different heights. Many others live in stems hollowed out by beetles.

construction of roads, the improvement of waterways, and the provision of electricity from hydroelectric dams. Brazilians are just now learning about what is being lost and conservation efforts are growing. Among areas set aside are Chapada de Guimarães National Park, Emas National Park, and Bodoquena National Park bordering the Pantanal and Serra de Canastra and Serra do Cipó National Parks in the southeast. However, the lands surrounding these parks are being cleared of natural vegetation, leaving smaller and smaller fragments of undisturbed cerrado. Plans exist to create ecological corridors that will connect cerrado patches and gallery forests and permit the movement of animals and plants that is necessary if a given site is to recover after disturbance and to slow the loss of biodiversity caused by decreasing habitat area and increasing habitat fragmentation.

Further Readings

Book

Mistry, J. 2000. *World Savannas: Ecology and Human Use*. Harlow, England: Pearson Education Ltd.

Internet Sources

African Wildlife and Conservation. n.d. www.wildwatch.com.

The Northlands Dung Beetle Express. n.d. "Dung beetle biology." www.dungbeetles.com.au/index.pl.

PlantZAfrica. n.d. www.plantzafrica.com.

WildWorld. n.d. www.worldwildlife.org/wildworld/profiles.

Climate Data

WorldClimate. 1996–2008. World Climate. www.worldclimate.com.

Representative Stations:

Africa

- East African savanna: Nairobi, Kenya
- South African savannas
  - Kalahari dry savanna: Upington, South Africa
  - Moist savanna (vicinity Kruger National Park): Nelspruit, South Africa
- West African savannas
  - Guinean zone: Odienne, Ivory Coast
  - Sahelian zone: Ndjamena, Chad
  - Sudanian zone: Kano, Nigeria

Australia

- Savanna woodland: Katherine, Australia

South America

- Cerrado: Cuiaba, Brazil
- Llanos: Cuidad Bolivar, Venezuela

Appendix

Selected Plants and Animals of the Tropical Savanna Biome

African Savannas

Some Characteristic Plants of the Sahelian Zone of the West African Savanna

Annual grasses	
Indian sandbur	*Cenchrus biflorus*
Grass	*Schoenefeldia gracilis*
Grass	*Aristida stipoides*
Woody plants	
Acacia/Umbrella tree	*Acacia tortilis*
Acacia	*Acacia laeta*
Acacia	*Acacia ehrenbergiana*
Corkwood	*Commiphora africana*
Desert date	*Balanites aegyptiaca*
Evergreen shrub	*Boscia senegalensis*

Some Characteristic Animals of the Sahelian Zone of the West African Savanna

Large mammals	
Scimitar-horned oryx	*Oryx dammah*[a]
Dama gazelle	*Gazella dama*
Dorcas gazelle	*Gazella dorcas*
Red-fronted gazelle	*Gazella rufifrons*
Bubal hartebeest	*Alcelaphus busephalus busephalus*[b]
Wild dog	*Lycaon pictus*[c]
Cheetah	*Acinomyx jubatos*[c]
Lion	*Pantera leo*[c]

Small mammals[d]	
Gerbil	*Gerbillus bottai*
Gerbil	*Gerbillus muriculus*
Gerbil	*Gerbillus nancillus*
Gerbil	*Gerbillus stigmonyx*
Gerbil	*Taterillus petteri*
Gerbil	*Taterillus pygargus*
Zebra mouse	*Lemniscomys hoogstraali*

Notes: [a]Presumed extinct in the wild; [b]Extinct subspecies; [c]Largely extirpated throughout region; [d]Endemic to the Sahel region.

Some Characteristic Plants of the Sudanian Zone of the West African Savanna

Perennial grasses	
Elephant grass	*Pennisetum purpureum*
Elephant or thatching grass	*Hyparrhenia* spp.
Woody plants	
Isoberlinias	*Isoberlinia* spp.
Terminalias	*Terminalia* spp.
Combretums/bushwillows	*Combretum* spp.
Acacia	*Acacia seyal*
Acacia	*Acacia albida*
Acacia	*Acacia nilotica*
Baobab	*Adonsonia digitata*

Some Characteristic Large Mammals of the Sudanian Zone of the West African Savanna

Herbivores	
Black rhinoceros	*Diceros bicornis*
Northern white rhinoceros	*Ceratotherium simum cottoni*
Elephant	*Loxodonta africanus*
Western giraffe	*Giraffa camelopardalis peralta*
West African savanna buffalo	*Syncerus caffer brachyceros*
Western giant eland	*Taurotragus derbianus derbianus*
Roan antelope	*Hippotragus equinus*
Carnivores	
Wild dog	*Lycaon pictus*
Cheetah	*Acinomyx jubatus*
Leopard	*Panthera pardus*
Lion	*Panthera leo*

Some Characteristic Plants of the Guinean Zone of the West African Savanna

Perennial grasses	
Elephant grass or thatching grass	*Hyparrhenia rufa*
Bluestem	*Andropogon* spp.
Panicum	*Panicum* spp.
Lemongrass	*Cymbopogon* spp.
Balsamscale	*Elionurus* spp.
Woody plants of the savanna woodlands	
Isoberlinia	*Isoberlinia doka*
Isoberlinia	*Isoberlinia tomentosa*
Terminalia	*Terminalia avicenniodes*
Parkia	*Parkia biglobosa*
African mesquite	*Prosopis africana*
Tree	*Anogeissus leiocarpus*
Tree	*Lophira lancelota*
Baobab	*Adansonia digitata*
Trees of the gallery forests	
Lingue	*Afzelia africana*
Kapok tree	*Ceiba pentandra*
Raphia palm	*Raphia sudanica*
African oil palm	*Elaeis guineensis*

Some Characteristic Animals of the Guinean Zone of the West African Savanna

Mammals	
Herbivores	
Black rhinoceros	*Diceros bicornis*
Elephant	*Loxodonta africana*
Hippopotamus	*Hippopotamus amphibius*
Warthog	*Phacochorerus aethiopicus*
Red-fronted gazelle	*Gazella rufifrons*
Giant eland	*Taurotragus derbianus*
Roan antelope	*Hippotragus equinus*
Kob	*Kobus kob*
Waterbuck	*Kobus ellipsiprymnus*
Topi	*Damaliscus lunatus*
Reedbuck	*Redunca redunca*
Buffalo	*Syncerus cafer*
Baboon	*Papio papio*
Colobus monkey	*Colobus guereza*
Patas monkey	*Cercopithecus patas*

Carnivores	
Leopard	*Panthera pardus*
Cheetah	*Acinonyx jubatus*
Wild dog	*Lycaon pictus*
Birds	
Ostrich	*Struthio camelus*
Shoebill	*Balaeniceps rex*
Marabou Stork	*Leptoptilos crumeniferus*
Reptile	
Nile crocodile	*Crocodylus niloticus*

Some Characteristic Plants of the East African Savannas

Grasses	
Red grass	*Themada trianda*
Purple pigeon grass	*Setaria incrassata*
Kleingrass	*Panicum coloratum*
Common needlegrass/Sixweek three awn	*Aristida adscensionis*
Lovegrasses	*Eragrostis* spp.
Bluestems	*Andropogon* spp.
Woody plants	
Acacia/Umbrella tree	*Acacia tortilis*
Corkwoods	*Commiphora* spp.
Grewia	*Grewia bicolor*
Terminalia	*Terminalia spinosa*
Baobab	*Adansonia digitata*

Some Characteristic Animals of the Eastern and Southern African Savannas

Mammals	
Herbivores	
African elephant	*Loxodonta africana*
White rhinoceros	*Ceratotherium simum*
Black rhinoceros	*Diceros bicornis*
Plains zebra	*Equus burchelli*
Grevy's zebra	*Equus grevyi*
Warthog	*Phacochorus aethiopicus*
Giraffe	*Giraffa camelopardis*
African buffalo	*Syncerus cafer*
Blue wildebeest or gnu	*Connochaetes taurinus*
Common hartebeest	*Alcephalus busephalus*

(*Continued*)

Eland	*Taurotragus oryx*
Topi	*Damaliscus lunatus*
Thomson's gazelle	*Gazella thomsoni*
Dama gazelle	*Gazella dama*
Springbok	*Antidorca marsupalis*
Gerenuk	*Litocranius walleri*
Impala	*Aepyceros melampus*
Dikdik	*Madoqua kirki*
Oryx	*Oryx gazella*
Insect eaters	
Hedgehog	*Erinaceus albiventris*
Elephant shrew	*Elephatus rufescens*
Common pangolin	*Manis temminckii*
Aardvark	*Orycteropus afer*
Banded mongooses	*Mungos mungo*
Aardwolf	*Proteles cristatus*
Carnivores	
Lion	*Panthera leo*
Leopard	*Panthera pardus*
Cheetah	*Acinonyx jubatus*
Caracal	*Felis caracal*
Serval	*Felis serval*
Hunting dog	*Lysoan pictus*
Common or golden jackal	*Canis aureus*
Spotted hyena	*Crocuta crocuta*
Brown hyena	*Hyaena brunnea*
Striped hyena	*Hyaena hyaena*
Omnivores	
Honey badger	*Mellivera capensis*
African civet	*Vivera civetta*
Savanna baboon	*Papio cynocephalus*
Birds	
White-headed Vulture	*Trigonoceps occipitalis*
Rüppell's Griffon	*Gyps rueppellii*
Lappet-faced Vulture	*Aegypius tracheliotus*
Ostrich	*Struthio camelus*
Secretary Bird	*Sagittarius serpentarius*
Kori Bustard	*Ardeotis kori*
Social Weaver	*Philetirus socius*
Red-billed Buffalo-Weaver	*Bubalornis niger*
Red-billed Quelea	*Quelea quelea*
Invertebrates	
Termites	*Macrotermes* spp.

Some Characteristic Grasses of the Serengeti Plains

Mid-size grasses	
Red grass	*Themada trianda*
Common russet grass	*Loudetia simplex*
Yellow thatching grasses	*Hyperthelia dissoluta*
Bamboo grass/Rabbit tail fountain grass	*Pennisetum mezianium*
Short grasses	
Bluestem	*Andropogon greenwayi*
Dropseeds	*Sporobolus ioclados, Sporobolus kentrophyllus, Sporobolus spicatus*
Kleingrass	*Panicum coloratum*
Rhodes grass	*Chloris gayana*
Fingergrass	*Digitaria macroblephara*
Bermuda grass	*Cynadon dactylon*

Some Characteristic Plants of the Kalahari

Annual grasses	
Kalahari sourgrass	*Schmidtia kalihariensis*
Nine-awned grass	*Enneapogon cenchroides*
Perennial grasses	
White buffalograss	*Panicum coloratum*
Tall bushman grass	*Stipagrostis cilite*
Small bushman grass	*Stipagrostis obtusa*
Silky bushman grass	*Stipagrostis uniplumis*
Lehamnn's lovegrass	*Eragrostis lehmanniana*
Trees	
Camel thorn	*Acacia erioloba*
Grey camel thorn	*Acacia haematoxylon*
False umbrella thorn	*Acacia luederitzii*
Wild green hair tree	*Parkinsonia africana*
Shepherd's tree	*Boscia albitrura*
Silver clusterleaf	*Terminalea sericea*
Shrubs	
Blackthorn	*Acacia mellifera*
Blue pea	*Lebeckia linearifolia*
Velvet raisin	*Grewia flava*
Succulents	
Wild ghaap	*Hoodia gordonii*
Ouheip	*Adenium oleifolius*

(*Continued*)

Bulb	
Vlei lily	*Nerine lacticoma*
Creepers	
Tsamma melon	*Citrullus lanatus*
Devil's thorn	*Tribulus terrestris* and *Tribulus zeyherie*
Annual forbs	
Yellow mouse whiskers or pretty lady	*Cleome angustifolia*
Wild everlastings	*Helichrysum argyrosphaerum*
Thunderbolt flower	*Sesamum triphyllum*

Some Characteristic Animals of the Kalahari

Mammals	
Herbivores	
Porcupine	*Hystrix africeaustralis*
Ground squirrel	*Xerus inauris*
Striped mouse	*Rhabdomys pumilo*
Whistling rat	*Paratomys brantsii*
Springhare	*Pedetes capensis*
Steenbok	*Raphicerus campestis*
Springbok	*Antidorcas marsupalis*
Gemsbok	*Oryx gazella*
Kudu	*Tragelaphus strepsiceros*
Red hartebeest	*Alcephalus buselaphus*
Blue wildebeest	*Connochaetes taurinus*
Carnivores	
Yellow mongoose	*Cynictis penicillata*
Honey badger	*Mellivora capensis*
Cape fox	*Vulpes chama*
Bat-eared fox	*Otocyon megalotis*
Black-backed jackal	*Canis mesomelas*
Brown hyena	*Hyaena brunnea*
Meercat (suricate)	*Suricata suricata*
African wild cat	*Felis silvestris*
Caracal	*Felis caracal*
Lion	*Panthera leo*
Leopard	*Panthera pardalis*
Cheetah	*Acinonyx jubotus*
Aardvark or antbear	*Oryctreopus afer*
Birds	
Ostrich	*Struthio camelus*
Southern Pale Chanting Goshawk	*Melierax canorus*
Bateleur	*Terathopius ecaudatus*
Lappet-faced Vulture	*Torgos tracheliotus*

White-backed Vulture	*Gyps africanus*
Spotted Eagle Owl	*Bubo africanus*
Kori Bustard	*Ardeotis kori*
Secretary Bird	*Sagittarius serpentarius*
Swallow-tailed Bee-eater	*Merops hirundineus*
Southern Yellow-billed Hornbill	*Tockus leucomelas*
Fork-tailed Drongo	*Dicrurus adsimilis*
White-browed Sparrow-Weaver	*Plocepasser mahali*
Sociable Weaver	*Philetairus socius*
Red-billed Quelea	*Quelea quelea*
Shaft-tailed Whydah	*Vidua regia*
Reptiles	
Leopard tortoise	*Geochelone pardalis*
Ground agama	*Agama aculeata*
Common barking gecko	*Ptenopus garralus*

Some Characteristic Plants of Moist Savanna in Kruger National Park, South Africa

Trees of the bushveld	
Knobthorn acacia	*Acacia nigrens*
Umbrella acacia	*Acacia tortilis*
Sweet-thorn acacia	*Acacia karroo*
Silver clusterleaf	*Terminalis sericea*
Magic guarri	*Euclea divinorum*
Red bushwillow	*Combretum apiculatum*
Sicklebush	*Dischrostachys cinereus*
Marula	*Sclerocarya birrea*
Sausage tree	*Kigelia africana*
Trees of mopane savanna	
Mopane	*Colosphospermum mopane*
Red bushwillow	*Combretum apiculatum*
Purple-pod clusterleaf	*Terminalia prunoides*
Magic guarri	*Euclea divinorum*
Fever tree	*Acacia xanthophloea*
Sycamore fig	*Ficus sycamorus*
Natal mahogany	*Trichilia emetica*
Baobab	*Adansonia digitata*
Tall bunchgrasses	
Tamboekie grass	*Hypertheliea dissoluta*
Thatching grass	*Hyperthelia filipendula*
Red grass	*Themeda trianda*
Natal red-top grass	*Rhynchelytrem repens*
Guinea or buffalo grass	*Panicum maximum*

Some Characteristic Animals of the Moist Savannas of Southern Africa

Mammals	
Herbivores	
Elephant	*Loxodonta africanus*
Plains or Burchell's zebra	*Equus burchelli*
White or square-lipped rhinoceros	*Ceratotherium simum*
Black or hook-lipped rhinoceros	*Diceros bicornis*
Hippopotamus	*Hippopotamus amphibious*
Warthog	*Phacochoerus africanus*
Giraffe	*Giraffa camelopardis*
Buffalo	*Syncerus caffer*
Eland	*Taurotragus oryx*
Blue wildebeest	*Connochaetes taurinus*
Kudu	*Tragelaphus strepsiceros*
Nyala	*Tragelaphus angasi*
Bushbuck	*Tragelaphus scriptus*
Waterbuck	*Kobus ellipsiprymnus*
Impala	*Aepyceros melampus*
Tsessebe	*Damaliscus lunatus*
Steenbok	*Raphiceros campestris*
Tree squirrel	*Paraxerus cepapi*
Carnivores	
Aardvark	*Orycteropus afer*
Wild dog	*Lycaon pictus*
Black-backed jackal	*Canis mesomelas*
Side-striped jackal	*Canis adustus*
Spotted hyena	*Crocuta crocuta*
Civet	*Civettictis civetta*
Dwarf mongoose	*Helogale parvula*
Lion	*Panthera leo*
Leopard	*Panthera pardalis*
Cheetah	*Acinonyx jubotus*
Omnivores	
Savanna or chacma baboon	*Papio cynocephalus*
Vervet monkey	*Chlorocebus aethiops*
Birds	
Three-banded Plover	*Charadrius tricollaris*
Crowned Lapwing	*Vanellus coronatus*
Water Thick-knee	*Burhinus vermiculatus*
Hamerkop	*Scopus umbretta*
African Fish-Eagle	*Haliaeetus vocifer*
Brown Snake Eagle	*Circaetus cinereus*

Lappet-faced Vulture	*Torgos tracheliotus*
Pied Kingfisher	*Ceryle rudis*
Saddleback Stork	*Ephippiorrhyncus senegalensis*
Marabou Stork	*Leptotilos crumeniferus*
Red-breasted Korhaan	*Eupodotis ruficusta*
Helmeted Guineafowl	*Numidia meleagris*
Swainson's Spurfowl or Francolin	*Pternistis swainsonii*
Southern Ground Hornbill	*Bucorvus leadbeateri*
African Grey Hornbill	*Tockus nasutus*
Red-billed Oxpecker	*Buphagus erythrorhynchus*
Carmine Bee-eater	*Merops nubicoides*
Lilac-breasted Roller	*Coracias caudatus*
Go-away Bird (Lorie)	*Corythaixoides concolor*
Magpie Shrike	*Corvinella melanoleuca*
Burchell's Starling	*Lamprotornis australis*
Red-winged Starling	*Onychognathus morio*
Southern Masked Weaver	*Ploceus velatus*
Red-billed Quelea	*Quelea quelea*
Reptiles	
Leopard tortoise	*Geochelone pardalis*
Hinged tortoise	*Kinixys* sp.
Nile crocodile	*Crocodylus niloticus*

Australian Savannas

Some Characteristic Plants of Australia's Northern Monsoon Tallgrass Savanna

Grasses	
Kangaroo grass	*Themeda triandra*
Wild sorghums	*Sorghum* spp.
Senale redgrass	*Schizachryium fragile*
Golden beardgrass	*Chrysopogon fallax*
Woody plants	
Wattle	*Acacia* spp.
Gums/Eucalypts	*Eucalyptus tetrodonta, E. miniata*
Bloodwood	*Eucalyptus terminalis*
Paperbark trees	*Melaleuca* spp.
Hakeas	*Hakea* spp.
Banskias	*Banksia* spp.
Bean or orchid tree	*Bauhinia* spp.
Terminalia	*Terminalia* spp.

Some Characteristic Plants of Australia's Southern Tropical and Subtropical Tallgrass Savannas

Grasses	
Black spear grass	*Heteropogon contortus, Heteropogon triticean*
Kangaroo grasses	*Themeda* spp.
Bluestems	*Bothriochloa* spp.
Woody plants	
Wattles	*Acacia* spp.
Gums/Eucalypts	*Eucalyptus* spp.
Paperbark tree	*Melaleuca quinquenervia*
Hakeas	*Hakea* spp.
Banskias	*Banksia robur*
Bean or orchid tree	*Bauhinia* spp.
Terminalia	*Terminalia* spp.

Some Characteristic Plants of Australia's Midgrass Savannas

Grasses	
Three-awns	*Aristida* spp.
Queensland bluegrass	*Dichanthium sericeum*
Redleg grass	*Bothriochloa decipiens*
Forest blue grass	*Bothriochloa bladhii*
Windmill grasses	*Chloris* spp.
Woody plants	
Wattles	*Acacia* spp.
Gums/Eucalypts	*Eucalyptus* spp.
Paperbark trees	*Melaleuca* spp.
Hakeas	*Hakea* spp.
Banskias	*Banksia* spp.
Bean or orchid tree	*Bauhinia* spp.
Terminalia	*Terminalia* spp.

Some Characteristic Plants of Australia's Mulga Pastures

Grasses	
Crab grasses	*Digitaria* spp.
Bandicoot grass	*Monochatha paradoxa*
Eriachne grasses	*Eriachne* spp.
Three-awns	*Aristida* spp.
Woody plant	
Mulga	*Acacia aneura*

Some Characteristic Animals of Australia's Tropical Savannas

Mammals	
Marsupials	
Grey kangaroo	*Macropus giganteus*
Agile wallaby	*Macropus agilis*
Antilopine wallaroo	*Macropus antilpinus*
Northern brown bandicoot	*Isoodon macrurus*
Northern quoll	*Dasyrus halluctus*
Brushtail possum	*Trichosurus vulpecula*
Sugar glider	*Petaurus breviceps*
Placentals	
Rats	*Rattus* spp.
Banana rats	*Melomys* spp.
Tree rats	*Colinurus* spp.
Water rat	*Hydromys* spp.
Birds	
Emu	*Dromaius novaehollandiae*
Lorikeets	several genera
Cockatoos	several genera
Parrots	several genera
Honeyeaters	family Meliphagidae
Butcherbirds	family Artamidae
Reptiles	
Frill-necked lizard	*Chlamydosaurus kingii*
Sand goanna	*Voranus gouldii*
Two-lined dragon	*Diphorophora billenata*

South American Savannas

Some Characteristic Trees of Gallery Forests in the Llanos

Trees	
Llanos palm	*Copernicia tectorum*
Monkey-pod	*Pithecellobium saman*
Figs	*Ficus* spp.
Genipap	*Genipa americana*
Palo de agua	*Cordia collococa*
Panama tree	*Sterculia apetala*
Kapok tree	*Ceiba pentandra*
Verdolago	*Terminalia amazonica*
Mahogany	*Swietenia macrophylla*

Some Characteristic Plants of the Llanos: Bajiós

Grasses	
Bamboo grass	*Hymenachne amplexicaulis*
Cut-grass	*Leersia hexandra*
Paspalums	*Paspalum* spp.
Spikerushes	*Eleocharis* spp.
Trachypogons	*Trachypogon* spp.
Palm	
Moriche palm	*Mauritia flexuosa*
Spiny shrubs	
Barinas	*Cassia aculeata*
Mimosas	*Mimosa pigra, Mimosa dormiens*
Guaica	*Randia armata*

Some Characteristic Plants of the Llanos: Esteros

Floating plants	
Water hyacinths	*Eichhornia crassipes, Eichhornia azurea*
Water ferns	*Salvinia* spp.
Water lettuce	*Pistia stratiotes*
Water primroses	*Ludwigia* spp.
Rooted plants	
Fire flag	*Thalia geniculata*
Tropical morning glories	*Ipomoea crassicaulis, Ipomoea fistulosa*
Spikerushes	*Eleocharis* spp.
Sedges	*Cyperus* spp.

Some Characteristic Plants of the Llanos: High Plains and High Plains

Grasses	
Trachypogons	*Trachypogon plumosus, Trachypogon vestitus*
Bluestems	*Andropogon semiberis, Andropogon selloanus*
Carpetgrass	*Axonopus anceps*
Paspalum	*Paspalum carinatum*
Dropseed	*Sporobolus indicus*
Woody plants	
Nance or manteco	*Brysonima crassifolia*
Sandpaper tree or chaparro	*Curatella americana*
Sucopira or alconque	*Bowdichia virgiloides*
Palms	
Moriche palm	*Mauritia flexuosa*

Some Characteristic Animals of the Llanos

Mammals	
Herbivores	
Giant anteater	*Myrmecophaga tridactyla*
Southern tamandua	*Tamandua tetradactyla*
Llanos long-nosed armadillo	*Dasypus sabanicola*
White-tailed deer	*Odocoileus virginianus*
Capybara	*Hydrochoerus hydrochaeris*
Carnivores	
Giant otter	*Pteronura brasiliensis*
Crab-eating fox	*Cercocyon thous*
Ocelot	*Leopardus pardalis*
Puma	*Felis concolor*
Jaguar	*Panthera onca*
Reptiles	
Orinoco crocodile	*Crocodylus intermedius*
Spectacled caiman	*Caiman crocodilus*
Savanna side-necked turtle	*Podocnemis vogli*
Green iguana	*Iguana iguana*
Tegu lizard	*Turpinambis teguixin*
Giant green anaconda	*Eunectes murinus*
Amphibians	
Cane toad	*Bufo marinu*
Emerald-eyed treefrog	*Hyla crepitans*
Yellow treefrog	*Hyla microcephala*
Rivero's tiny treefrog	*Hyla miniscula*
Caracas snouted treefrog	*Scinax rostrata*
Colombian four-eyed frog	*Pleurodema brachyops*
Swimming frog	*Pseudis paradoxa*

Some Characteristic Plants of the Cerrados

Grasses	
Bunchgrass	*Tristachya chrysothrix*
Goat's beard	*Aristida pallens*
Bunchgrass	*Leptocorphyium lanatum*
Bunchgrass	*Trachypogon spicatus*
Subshrubs	
Wild custard apple	*Annona pygmaea*
Catuaba	*Anemopaegma arvense*
Cabbage bark	*Andira humilis*

(*Continued*)

Monkey-nut	*Anacardium humile*
Arnica	*Lychnophora ericoides*
Arnica	*Pseudobrickellia pinifolia*
Rabo de tatu	*Centrosema bracteosum*
Trees	
Pequi	*Caryocar brasiliensis*
Murici	*Byrsonima coccolobifolia*
Nance	*Byrsonima crassifolia*
Lixeiro or sandpaper tree	*Curatella americana*
Barbatimão	*Stryphnodendron barbatimao*
False barbatimão	*Dimorphandra mollis*
Pau santo	*Kielmeyera coriacea*
Pau de arara	*Salvertia convallaroides*
Pau terra	*Qualea grandifolia*
Pau terra	*Qualea parvifolia*
Pau terra	*Voschrysia thyrsoidea*

Common Plants of Wet Areas in the Cerrado

Stoloniferous grasses	
	Axonopus purpusii
	Paspalum morichalense
	Reimarochloa acuta
Terrestrial orchids	
	Habenaria (more than 70 species)
Palms	
Buriti	*Mauritia vinifera*
Buritirana	*Mauritia martiana*
Woody plants of earthmounds (murundus)	
Lixeiro	*Curatella americana*
Nance	*Byrsonima crassifolia*
Cabbage bark	*Andira cuiabensis*

Some Characteristic Animals of the Cerrados

Mammals	
Herbivores	
White-eared opossum	*Didelphis albiventris*
Giant armadillo	*Priodontes maximus*
Southern tamandua or lesser anteater	*Tamandua tetradactyla*
Giant anteater	*Myrmecophaga tridactyla*
Black tufted-ear marmoset	*Callithrix penicillata*

South American tapir	*Tapirus terrestris*
Collared peccary	*Tayassu tajucu*
White-lipped peccary	*Tayassu pecari*
Brocket deer	*Mazama americana, Mazama gouazoupira*
Pampas deer	*Ozotoceros bezoarcticus*
Marsh deer	*Blastoceros dichotomus*
Paca	*Agouti paca*
Agouti	*Dasyprocta agouti*
Brazilian guinea pig	*Cavia aperea*
Prehensile-tailed porcupine	*Coendou prehensilis*
Carnivores	
Maned wolf	*Chrysocyon brachyurus*
Jaguarundi	*Herpailurus yagouaroundi*
Jaguar	*Panthera onca*
Birds	
Rhea or Ema	*Rhea americana*
Hyacinth Macaw	*Anodorhynchus hyacinthinus*
Blue-fronted Amazon	*Amazona aestiva*
Yellow-faced Amazon	*Amazona xanthops*
Blue and Gold Macaw	*Ara ararauna*
Peach-fronted Parakeet	*Aratinga aurea*
Blue-headed Parrot	*Pionus menstrus*
Small-billed Tinamou	*Crypturellus parvirostris*
Red-winged Tinamou	*Rhynchotus rufescens*
Spotted Tinamou	*Nothura maculosa*
Red-legged Seriema	*Cariama cristata*
Burrowing Owl	*Athene cunicularia*
Crested Caracara	*Polyborus plancus*
Lesser Yellow-headed Vulture	*Cathartes burrovianus*
Campo Flicker	*Colaptes campestris*
White-collared Swift	*Streptoprocne zonaris*
Reptiles	
Red-footed tortoise	*Geochelone carbonaria*
Boa constrictor	*Boa constrictor*
False coral snake	*Erythrolamprus aesculapii*
Brazilian rattlesnake	*Crotalus durissus*
Coral snake	*Micrurus frontalis*
False chameleon	*Polychrus acutirostris*
Iguanid lizard	*Tropidurus torquatus*
Giant worm lizard	*Amphisbaena alba*

(*Continued*)

Invertebrates	
Neotropical termite	*Armitermes euamignathus*
Neotropical termite	*Cornitermes snyderii*
Leaf-cutter ants	*Atta* spp. and *Acromyrmex* spp.

Glossary

Adaptation. Any inherited aspect of morphology, physiology, or behavior that improves a species chances of long-term survival and reproductive success in a particular environment.

Annual. A plant that completes its life cycle from seed to mature reproducing individual to death in a year's time or less.

Artiodactyl. Any of the even-toed ungulates, including antelopes, bison, and giraffes. With digestive systems geared to the fermentation of grasses, these mammals came to dominate the world's grasslands.

Biodiversity. The total variation and variability of life in a given region. Determined at various scales, including genes, species, and ecosystems.

Biome. A large subcontinental region of similar vegetation, animal life, and soils adapted to the area's physical environmental conditions such as climate or some disturbance factor.

Browser. A plant-eating animal that specializes on leaves, stems, or twigs. Compare with grazer.

Bulrush. A grass-like plant (graminoid) in the genus *Scirpus*. Associated with wetlands, it bears clusters of small brown florets.

C_3 Grass. A cool-season grass that uses the chemical pathways of the Calvin Cycle to assimilate carbon dioxide into 3-carbon molecules during photosynthesis. Most green plants fall into this category.

C_4 Grass. A warm-season grass that uses 4-carbon molecules to convert carbon dioxide to organic molecules during photosynthesis. Common among tropical grasses.

Cape Floristic Province. Another term for the South African Floristic Kingdom, a small area surrounding the Cape of Good Hope that is renowned for its extremely high diversity of plant species.

Caviomorph (rodent). One of a group of distinctly South American rodents that includes the cavies, capybara, and mara. The porcupine is the only caviomorph in North America.

Climate. The general weather patterns expected during an average year. The main factors are temperature and precipitation. See also weather.

Cold-Season Grass. Grasses that grow best at moderate temperatures and often become dormant during the hottest part of the year, resuming growth in early spring when the lower temperatures of autumn arrive. These are C_3 grasses typical of the temperate grasslands.

Culm. The stem of a grass or other graminoid.

Cushion Plant. A low, many-stemmed plant that grows as a dense mound.

Disturbance. An event that disrupts an ecosystem and damages or destroys some part of it. Origin may be biological (such as overgrazing) or physical (such as edaphic conditions, fire, or flood).

Domesticated. Produced by the selective breeding of humans or by natural adaptation to human-dominated environments.

Edaphic. Pertaining to conditions of the substrate or soil. Those that affect plant growth include nutrient depletion, poor drainage, excessive drainage, and presence of a hardpan.

El Niño. A seasonal weather phenomenon that affects the equatorial Pacific, especially off the west coast of South America. During these events of December, normal high-pressure systems that make the coast exceptionally dry are replaced by low pressure, high humidity, and even rain. Severe, prolonged El Niños can affect weather patterns around the world.

Endemic. Native to and restricted to a particular geographic area.

Exotic. Nonnative, introduced, alien.

Fauna. All the animal species in a given area, or some subset of them such as the bird fauna or grazing fauna.

Forb. An herb with broad leaves and soft, nonwoody stem. Wildflowers are typical of this growthform.

Fossorial. Burrowing. Adapted to living underground.

Fynbos. The local name for mediterranean scrub vegetation in South Africa's Cape Floristic Province.

Graminoid. The growthform of grasses, sedges, rushes, and reeds. A type of herb.

Grass. Any flowering plant of the family Poaceae (also known as the family Graminae).

Grazer. A plant-eating animal that consumes primarily grasses. Compare with browser.

Hardpan. A dense rock-like layer of substrate which is difficult for water or roots to penetrate.

Herb. A nonwoody plant that dies down each year. May be an annual or a perennial. This growthform includes graminoids and forbs.

Herbaceous. Having the characteristics of an herb.

Honeydew. A sweet, sticky substance secreted by aphids and some scale insects

Hotspot. An area of high biodiversity, usually threatened and requiring protection.

Humus. Well-decayed plant matter that is collidal in size and assumes a dark brown color. Because it helps hold moisture and nutrients in a soil, humus content is an indicator of soil fertility.

Ion. A particle bearing a negative or positive charge.

ITCZ (Intertropical Convergence Zone). The contact zone between the Trade Winds of the Northern and Southern Hemispheres. Shifts its position north and south of the equator with the seasons and, when overhead, usually brings rain.

Lagomorph. A rabbit, hare, or pika. A member of the order Lagamorpha.

Laterite. A hardpan that forms in red tropical soils (oxisols) because of the concentration of iron oxide.

Latitude. The distance of a point north or south of the equator (0° Latitude), measured in degrees.

Leaching. The process in which dissolved substances are removed from the substrate by the downward percolation of water.

Legume. Any plant that is a member of the pea and bean family (Leguminoseae) or, in other taxonomic schemes, Mimosaceae, Caesalpinaceae, or Fabaceae. Many of these plants have symbiotic relationships with rhizobial bacteria on their roots that fix atmospheric nitrogen into nitrates, important plant nutrients.

Loess. Powder-like material deposited by the wind. The most fertile temperate soils, the chernozems, are generally developed on this type of substrate.

Marsupial. A nonplacental mammal of the order Marsupalia whose young develop attached to teats in a pouch. Considered an early or primitive form of mammal, marsupials dominate the mammalian fauna of Australia and are also diverse in the neotropics.

Mediterranean. Refers to regions or climate patterns where winter is the rainy season and summers are dry.

Migration. The seasonal movement of a population from one area to another with a return to the original range when seasons change.

Monsoon. A wind that reverses its direction seasonally. An onshore flow typifies the warm season and an offshore flow occurs during the cold season. The Asian monsoon is most powerful and dominates the climate of the vast Indian Ocean region. More localized monsoonal systems occur elsewhere, as in the American Southwest.

Neotropical. Pertaining to the region from southern Mexico and the Caribbean to southern South America or to animals and plants restricted or nearly restricted to this region.

Nitrogen-Fixing. The ability to convert elemental or pure nitrogen to nitrates. Only certain micro-organisms can do this and hence they are vital to higher life forms that require nitrogen but cannot utilize it in its pure form. Rhizobial bacteria and cyanobacteria in the soil are important nitrogen-fixers in terrestrial ecosystems.

Perennial. A plant that lives two or more years.

Perissodactyl. Any of the odd-toed ungulates. Living forms include zebras, rhinoceroses, and tapirs.

Polar Front. In the global atmospheric circulation system, the contact zone between cold polar air and warmer subtropical air masses. Associated with uplift at the subpolar lows and often stormy, wet weather patterns in the mid-latitudes.

Rainshadow. A dry region that develops on the lee (downwind) side of mountain ranges. Low amounts of precipitation are the result of the warming of air masses as they descend the mountain slopes.

Ratite. A flightless bird lacking a broad sternum for the attachment of wing muscles. Large tropical grassland birds well adapted to running, ratites include ostriches, emus, cassowaries, and rheas.

Reed. A large hollow-stemmed grass of genera such as *Arundo* and *Phragmites* that possess plume-like inflorescences.

Rhizome. Underground stems from which new plantlets or tillers arise, leading to the sod-forming habit of some grasses.

Ruminant. An artiodactyl that possesses a four-chambered stomach in which grasses ferment and that chews its cud.

Sedge. A grass-like flowering plant in the family Cypercaceae. Members of the genus *Carex* usually prefer damper conditions than grasses.

Semiarid. Referring to climatic conditions in which there is too little precipitation to support forests, but not so little that deserts prevail. In the mid-latitudes, semiarid regions usually receive between 10 and 20 inches of precipitation a year and support natural grasslands.

Stolon. A horizontal stem that forms at ground level and gives rise to new tillers in grasses.

Stoma or Stomate. A tiny pore in the outer layer of a leaf through which gases, including water vapor, are exchanged with the atmosphere. The plural form is stomata.

Subshrub. A hard-stemmed shrub in which the upper branches die back during the nongrowing season. In a sense, a shrub that acts like a perennial forb.

Succulent. A growthform that permits the storage of water in some of its tissues. Plants may be leaf-succulents, stem succulents, or have special underground organs for storing water.

Tiller. A daughter plant or new plantlet forming from a grass's stolons or rhizomes.

Trade Winds. The strong, constant easterly winds of tropical latitudes.

Tropics. The latitudinal zone on Earth that lies between 23° 30 N and 23° 30 S; that is, between the Tropic of Cancer and the Tropic of Capricorn.

Tussock. A growthform of grasses and sedges in which individuals grow in clumps, forming visible hummocks.

Ungulate. A hoofed mammal.

Vegetation. The general plant cover of an area described in terms of its structure and appearance and not the species that comprise it.

Warm-Season Grass. A grass that photosynthesizes and grows best under high temperature conditions. These are C_4 grasses found in the tropics and in temperate grasslands with hot summers.

Weather. The state of the atmosphere at any given moment. Includes atmospheric pressure, temperature, humidity, and type of precipitation (if any).

Weathered. Pertaining to bedrock that has undergone physical or chemical breakdown into small particles, even ions.

Weed. A plant adapted to invade disturbed sites. Generally short-lived, they are good dispersers and fast growers.

Bibliography

General

Acton, D. F. 1992. "Grassland Soils." In *Natural Grasslands: Introduction and Western Hemisphere*, ed. R. T. Coupland, 25–54. Ecosystems of the World, 8A. Amsterdam: Elsevier.

Barbour, Michael G., and William Dwight Billings. 2000. *North American Terrestrial Vegetation.* 2nd ed. Cambridge: Cambridge University Press.

Bourliere, Francois, ed. 1983. *Tropical Savannas.* Ecosystems of the World, 13. Amsterdam: Elsevier.

Breckle, Siegmar-Walter. 2002. *Walter's Vegetation of the Earth. The Ecological Systems of the Geo-Biosphere.* 4th ed. Berlin: Springer Verlag.

Coupland, R. T., ed. 1992. *Natural Grasslands: Introduction and Western Hemisphere.* Ecosystems of the World, 8A. Amsterdam: Elsevier.

Coupland, R. T., ed. 1993. *Natural Grasslands: Eastern Hemisphere and Résumé.* Ecosystems of the World, 8B. Amsterdam: Elsevier.

Cowling, R. M., D. M. Richardson, and J. M. Pierce, eds. 1997. *Vegetation of Southern Africa.* Cambridge: Cambridge University Press.

Eyre, S. R. 1968. *Vegetation and Soils: A World Picture.* Rev. 2nd ed. Chicago: Aldine Publishing Company.

Furley, Peter A., and Walter W. Newey. 1983. *Geography of the Biosphere: An Introduction to the Nature, Distribution and Evolution of the World's Life Zones.* London: Buttersworth.

National Geographic and World Wildlife Fund. 2001. "Wild World, Terrestrial Ecoregions of the World." http://www.nationalgeographic.com/wildworld/terrestrial.html.

Nowak, Ronald M. 1991. *Walker's Mammals of the World.* 5th ed. 2 vols. Baltimore: Johns Hopkins University Press.

Oregon State University, Forage Information System (OSU/FIS). 2000. "Grass Growth and Regrowth for Improved Management." forages.oregonstate.edu/projects/regrowth.
Plant Conservation Alliance. Alien Plant Working Group. 2005. "Factsheet: Giant Reed." Weeds Gone Wild: Alien Plant Invaders of Natural Areas. www.nps.gov/plants/alien/fact/ardo1.htm.
Rinehart, Lee. 2006. "Switchgrass as a Bioenergy Crop." National Center for Appropriate Technology, National Sustainable Agriculture Information Service. http://attra.ncat.org/attra-pub/PDF/switchgrass.pdf.
USDA-NRCS PLANTS Database/Hitchcock, A. S. 1950. *Manual of Grasses of the United States*. Baton Rouge, LA: National Plant Data Center. http://plants.usda.gov.
Wild World. 2001. "Afrotropics: Tropical and Subtropical Grasslands, Savannas, and Shrublands." http://www.worldwildlife.org/wildworld/profiles/terrestrial_at.html#tropgrass.

Temperate Grasslands

North American

Coupland, R. T. 1992. "Mixed Prairie." In *Natural Grasslands: Introduction and Western Hemisphere,* ed. R. T. Coupland, 150–182. Ecosystems of the World, 8A. Amsterdam: Elsevier.
Daubenmire, Rexford. 1992. "Palouse Prairie." In *Natural Grasslands: Introduction and Western Hemisphere,* ed. R. T. Coupland, 297–312. Ecosystems of the World, 8A. Amsterdam: Elsevier.
Heady, H. F., J. W. Bartolome, M. D. Pitt, G. D. Savelle, and M. C. Stroud. 1992. "California Prairie." In *Natural Grasslands: Introduction and Western Hemisphere,* ed. R. T. Coupland, 313–335. Ecosystems of the World, 8A. Amsterdam: Elsevier.
Hoogland, John L. 2006. *Conservation of the Black-tailed Prairie Dog: Saving North American's Western Grasslands.* Washington, DC: Island Press.
Kucera, C. L. 1992. "Tall-Grass Prairie." In *Natural Grasslands: Introduction and Western Hemisphere,* ed. R. T. Coupland, 22–268. Ecosystems of the World, 8A. Amsterdam: Elsevier.
Laurenroth, W. K., and D. G. Milchunas. 1992. "Short-Grass Steppe." In *Natural Grasslands: Introduction and Western Hemisphere,* ed. R. T. Coupland, 183–226. Ecosystems of the World, 8A. Amsterdam: Elsevier.
Moul, Francis. 2006. *The National Grasslands.* Lincoln: University of Nebraska Press.
National Geographic and World Wildlife Fund. 2001. "Nearctic Temperate Grasslands, Savannas and Shrublands." Wild World, Terrestrial Ecoregions of the World. http://www.nationalgeographic.com/wildworld/profiles/terrestrial_na.html#tempgrassl.
Schmutz, E. M., E. L. Smith, P. R. Ogden, M. L. Cox, J. O. Klemmedson, J.J. Norris, and L. C. Fierro. 1992. "Desert Grassland." In *Natural Grasslands: Introduction and Western Hemisphere,* ed. R. T. Coupland, 337–362. Ecosystems of the World, 8A. Amsterdam: Elsevier.
United States Department of Agriculture Forest Service, Forest Health Staff. n.d. "Cheatgrass *Bromus tectorum*." Invasive Plants. http://www.na.fs.fed.us/fhp/invasive_plants/weeds/cheatgrass.pdf.
USDA-NRCS PLANTS Database/Hitchcock, A. S. 1950. *Manual of Grasses of the United States*. Baton Rouge, LA: National Plant Data Center. http://plants.usda.gov.

Vankat, John. 1979. *The Natural Vegetation of North America, an Introduction*. New York: John Wiley & Sons.

Weddell, Bertie J. 2001. "Fire in Steppe Vegetation of the Northern Intermountain Region." Idaho Bureau of Land Management, Technical Bulletin No. 01-14. http://www.id.blm.gov/techbuls/01_14/sec1.pdf.

Wind Cave National Park. n.d. "Grasses of the Mixed Grass Prairie." http://www.nps.gov/archive/wica/Grasses_of_the_Mixed_Grass_Prairie.htm.

Eurasian Steppe

Breckle, Siegmar-Walter. 2002. "Zonobiome of Steppes and Cold Deserts." In *Walter's Vegetation of the Earth. The Ecological Systems of the Geo-Biosphere,* ed. S.-W. Breckle, 371–416. 4th ed. Berlin: Springer Verlag.

Lavrenko, E. M., and Z. V. Karamysheva. 1993. "Steppes of the Former Soviet Union and Mongolia." In *Natural Grasslands: Eastern Hemisphere and Résumé,* ed. R. T. Coupland, 3–59. Ecosystems of the World, 8B Amsterdam: Elsevier.

Ting-Chen, Zhu. 1993. "Grasslands of China." In *Natural Grasslands: Eastern Hemisphere and Résumé,* ed. R. T. Coupland, 61–82. Ecosystems of the World, 8B Amsterdam: Elsevier.

Pampas

Coupland, R. T. 1992. "Overview of South American Grasslands." In *Natural Grasslands: Introduction and Western Hemisphere,* ed. R. T. Coupland, 363–366. Ecosystems of the World, 8A. Amsterdam: Elsevier.

Hawkins, H. S., and C. M. Donald. 1962. "Pasture Development in the Beef Cattle Regions of Argentina." *Grass and Forage Science* 17 (4): 245–259.

Pallares, Olegano Royo, Elbio J. Berreta, and Gerzy E. Maraschin. 2005. "The South American Campos Ecosystem." In *Grasslands of the World,* eds. J. M. Suttie, S. G. Reynolds, and C. Batello. Plant Production and Protection Series No. 34. Rome: Food and Agricultural Organization of the United Nations. www.fao.org//docrep/008/y8344e/y8344e0b.htm.

Patagonian Steppe

Darwin, Charles. 1962. *The Voyage of the Beagle.* Garden City, NY: Doubleday & Company, Inc.

Fernández, Robert J., and José M. Pareulo. 1988. "Root Systems of Two Patagonian Shrubs: A Quantitative Description Using a Geometrical Method." *Journal of Range Management* 41 (3): 220–223.

Veld

Low, A. Barrie, and A. G. Robelo, eds. 1996. "Vegetation of South Africa, Lesotho and Swaziland." Pretoria: Department of Environmental Affairs and Tourism. Grasslands Biome. http://www.ngo.grida.no/soesa/nsoer/Data/vegrsa/vegstart.htm.

O'Connor, T. G., and G. J. Bredenkamp. 1997. "Grassland." In *Vegetation of Southern Africa,* eds. R. M. Cowling, D. M. Richardson, and J. M. Pierce, 215–257. Cambridge: Cambridge University Press.

PlantZAfrica.com. n.d. "Grasslands Biome." South Africa National Biodiversity Institute. www.plantzafrica.com/frames/vegfram.htm.

South African National Parks. n.d. "Golden Gate Highlands National Park." www.sanparks.org/parks/golden_gate.

South African National Parks. n.d. "Mountain Zebra National Park." http://www.sanparks.org/parks/mountain_zebra.

Tainton, N. M., and B. H. Walker. 1993. "Grasslands of Southern Africa." In *Natural Grasslands: Eastern Hemisphere and Résumé,* ed. R. T. Coupland, 265–290. Ecosystems of the World, 8B. Amsterdam: Elsevier.

Tropical Savannas

General

Beard, J. S. 1953. "The Savanna Vegetation of Northern Tropical Amazonas." *Ecological Management* 23: 149–215.

Bourliere, Francois. 1983. "Mammals as Secondary Consumers in Savanna Ecosystems." In *Tropical Savannas,* ed. Francois Bourliere, 463–475. Ecosystems of the World, 13. Amsterdam: Elsevier.

Bourliere, F., and M. Hadley. 1983. "Present Day Savannas: An Overview." In *Tropical Savannas,* ed. Francois Bourliere, 1–17. Ecosystems of the World, 13. Amsterdam: Elsevier.

Breckle, Siegmar-Walter. 2002. "Zonobiome of Savannas and Deciduous Forests and Grasslands." In *Walter's Vegetation of the Earth. The Ecological Systems of the Geo-Biosphere.* ed. Siegmar-Walter Breckle. 4th ed. Berlin: Springer Verlag.

Cole, M. M. 1986. *The Savannas: Biogeography and Geobotany*. New York: Academic Press.

Fry, C. H. 1983. "Birds in Savanna Ecosystems." In *Tropical Savannas,* ed. Francois Bourliere, 337–357. Ecosystems of the World, 13. Amsterdam: Elsevier.

Happold, D. C. C. 1983. "Rodents and Lagomorphs." In *Tropical Savannas,* ed. Francois Bourliere, 363–400. Ecosystems of the World, 13. Amsterdam: Elsevier.

Hopkins, Brian. 1983. "The Soil Fauna of Tropical Savannas. III. The Termites." In *Tropical Savannas,* ed. Francois Bourliere, 505–524. Ecosystems of the World, 13. Amsterdam: Elsevier.

Institute of Terrestrial Ecology. 1996. "Lowland Tropical Grasslands." In Habitats of South America: Biotypes/Ecosystems Nomenclature, 144–154. http://www.naturalsciences.be/cb/ants/pdf_free/PHYSIS-HabitatsSouthAmerica.pdf.

Levieux, Jean. 1983. "Soil Fauna: The Ants." In *Tropical Savannas,* ed. Francois Bourliere, 525–540. Ecosystems of the World, 13. Amsterdam: Elsevier.

Mistry, J. 2000. *World Savannas: Ecology and Human Use*. Harlow, England: Pearson Education Ltd.

Montgomery, R. F., and G. P. Askew. 1983. "Soils of Tropical Savannas." In *Tropical Savannas,* ed. Francois Bourliere, 63–78. Ecosystems of the World, 13. Amsterdam: Elsevier.

Nix, H. A. 1983. "Climate of Tropical Savannas." In *Tropical Savannas,* ed. Francois Bourliere, 37–62. Ecosystems of the World, 13. Amsterdam: Elsevier.

Ojasti, J. 1983. "Ungulates and Large Rodents of South America." In *Tropical Savannas,* ed. Francois Bourliere, 427–439. Ecosystems of the World, 13. Amsterdam: Elsevier.

Rivas, J. A., J. V. Rodriguez, and G. G. Mittermeier. 2002. "The Llanos." In *Wildernesses,* ed. R. A. Mittermeier, 265–273. Mexico City: CEMEX. http://pages.prodigy.net/anaconda/llanos.htm.

Sarmiento, G., and M. Monasterio. 1983. "Life Forms and Phenology." In *Tropical Savannas,* ed. Francois Bourliere, 79–108. Ecosystems of the World, 13. Amsterdam: Elsevier.

Solbrig, O. T. 1996. "The Diversity of the Savanna Ecosystem." In *Biodiversity and Savanna Ecosystem Processes. A Global Perspective,* eds. O. T. Solbrig, E. Medina, and J. F. Silva, 1–127. Berlin: Springer.

African Savannas

African Wildlife and Conservation: Wildwatch. n.d. "Birds: Vultures." http://www.wildwatch.com/resources/birds/vultures.asp.

African Wildlife and Conservation: Wildwatch. n.d. "Plants: Baobab Tree—Africa's Oldest Inhabitants?" http://www.wildwatch.com/resources/plants/baobab.asp.

African Wildlife and Conservation: Wildwatch. n.d. "Termites: Towers of Clay." http://www.wildwatch.com/resources/other/termites.asp.

Anonymous. 2006. *The Kruger National Park Map Book.* Fourways, South Africa: Honeyguide Publications CC.

Cowling, R. M., D. M. Richardson, and J. M. Pierce, eds. 1997. *Vegetation of Southern Africa.* Cambridge: Cambridge University Press.

Duvall, Chris S. 2007. "Human Settlement and Baobab Distribution in South-Western Mali." *Journal of Biogeography* 34: 1947–1961.

Menaut, J. C. 1983. "African Savannas." In *Tropical Savannas,* ed. Francois Bourliere, 109–149. Ecosystems of the World, 13. Amsterdam: Elsevier.

Mobæk, Ragnhild, Anne Kjersti Narmo, and Stein R. Moe. 2005. "Termitaria Are Focal Feeding Sites for Large Ungulates in Lake Mburo National Park, Uganda." *Journal of Zoology, London* 267: 97–102.

Northlands Dung Beetle Express. n.d. "Dung Beetle Biology." www.dungbeetles.com.au/index.pl.

PlantZAfrica.com. n.d. "Adansonia digitata." http://www.plantzafrica.com/plantab/adansondigit.htm.

Sinclair, A. R. E. 1983. "The Adaptations of African Ungulates and Their Effects on Community Function." In *Tropical Savannas,* ed. Francois Bourliere, 4-1–426. Ecosystems of the World, 13. Amsterdam: Elsevier.

Sinclair, Ian, Phil Hockey, and Warwich Tarboten. 2002. *Birds of Southern Africa.* 3rd ed. Cape Town: Struik Publishers.

South African National Parks. 2004. *Kgalagadi National Park.* Pretoria: South African National Parks.

Stuart, Chris, and Tilde Stuart. 2001. *Field Guide to Mammals of Southern Africa.* 3rd ed. Cape Town: Struik Publishers.

Ulfstrand, Staffan 2002. *Savannah Lives: Animal Lives and Human Evolution in Africa.* Oxford: Oxford University Press.

United Nations Environmental Programme. 2003. "Comoé National Park, Côte d'Ivoire." www.unep-wcmc.org/sites/wh/comoe.html.

United Nations Environmental Programme. 1997. "Manovo-Gounda-St.Floris National Park." http://www.unep-wcmc.org/protected_areas/data/wh/manovo.html.

Van der Walt, Pieter, and Elias le Riche. 1999. *The Kalahari and Its Plants.* Pretoria: Published by the authors.

Wild World. n.d. "Serengeti Volcanic Grasslands (AT0714)." www.worldwildlife.org/wild world/profiles/terrestrial/at/at0714_full.html.

South American Savannas

Cartelle, Castor. 1999. "Pleistocene Mammals of the Cerrado and Caatinga of Brazil." In *The Central Neotropics: Ecuador, Peru, Bolivia, Brazil.* Vol. 3 of *Mammals of the Neotropics,* eds. John F. Eisenberg and Kent H. Redford, 27–46. Chicago: University of Chicago Press.

Fonseca, Gustavo A. B., Roberto B. Cavalcanti, Anthony B. Rylands, and Adriano P. Paglea. n.d. "Cerrado." www.biodiversityscience.org/publications/hotspots/Cerrado.html.

Furley, Peter A., and James A. Ratter. 1988. "Soil Resources and Plant Communities of the Central Brazilian Cerrado and Their Development." *Journal of Biogeography* 15 (1): 97–108.

Huber, Otto, Rodrigo Duno de Stefano, Gerardo Aymard, and Ricarda Riina. 2006. "Flora and Vegetation of the Venezuelan Llanos: A Review." In *Neotropical Savannas and Seasonally Dry Forests. Plant Diversity, Biogeography, and Conservation,* eds. R. Toby Pennington, Gwilym P. Leis, and James A. Ratter, 95–120. The Systematics Association Special Volume Series 69. New York: CRC, Taylor and Francis.

Kircher, John. 1997. *A Neotropical Companion.* 2nd ed. Princeton, NJ: Princeton University Press.

Nowak, Ronald M. 1991. *Walker's Mammals of the World.* 5th ed. 2 vols. Baltimore: Johns Hopkins University Press.

Ojasti, Juhani. 1991. "Human Exploitation of Capybara." In *Neotropical Wildlife Use and Conservation,* eds. John G. Robinson and Kent H. Redford, 236–252. Chicago: University of Chicago Press.

Oliveira, Paulo E., and Robert J. Marquis, eds. 2002. *The Cerrados of Brazil. Ecology and Natural History of a Neotropical Savanna.* New York: Columbia University Press.

Oliveira Filho, Ary T. 1992. "The Vegetation of Brazilian 'Murundus'—The Island Effect on the Plant Community." *Journal of Tropical Ecology* 8: 465–486.

Pinto, Maria Novaes, org. 1993. *Cerrado: Caracterizacão, Ocupacão e Perspectivas.* 2nd ed. Brasília: Editora Universidade de Brasília.

Rancy, Alceu. 1999. "Fossil Mammals of the Amazon as a Portrait of a Pleistocene Environment." In *The Central Neotropics: Ecuador, Peru, Bolivia, Brazil.* Vol. 3 of *Mammals of the Neotropics,* eds. John F. Eisenberg and Kent H. Redford, 3–26. Chicago: University of Chicago Press.

Ratter, James A., Samuel Bridgewater, and J. Felipe Ribeiro. 2006. "Biodiversity Patterns of the Woody Vegetation of the Brazilian Cerrado." In *Neotropical Savannas and Seasonally Dry Forests. Plant Diversity, Biogeography, and Conservation,* eds. Toby R. Pennington, Gwilym P. Leis, and James A. Ratter, 31–66. The Systematics Association Special Volume Series 69. New York: CRC, Taylor and Francis.

San Juan, A. 2005. "The Lurker's Guide to Leafcutter Ants." http://www.blueboard.com/leafcutters.

Sarmiento, G. 1983. "The Savannas of Tropical America." In *Tropical Savannas,* ed. Francois Bourliere, 245–288. Ecosystems of the World, 13. Amsterdam: Elsevier.

Simpson, George Gaylord. 1941. "The Vernacular Names of South American Mammals." *Journal of Mammalogy* 22 (1): 1–17.

United Nations Environmental Programme. n.d. "Cerrado Protected Areas: Chapada dos Veadeiros and Emas National Parks." www.unep-wcmc.org/sites/wh/cerrado.html.

Index

About the Author

SUSAN L. WOODWARD received her Ph.D. in geography from the University of California, Los Angeles, in 1976. She taught undergraduate courses in biogeography and physical geography for twenty-two years at Radford University in Virginia before retiring in 2006. Author of *Biomes of Earth*, published by Greenwood Press in 2003, she continues to learn and write about our natural environment. Her travels have allowed her to see firsthand some of the world's major grassland biomes in North America, South America, Russia, China, and southern Africa.